Calculations in Molecular Biology and Biotechnology

A Guide to Mathematics in the Laboratory

Frank H. Stephenson
Applied Biosystems

ACADEMIC PRESS

An imprint of Elsevier Science

Amsterdam Boston London New York Oxford Paris San Diego
San Francisco Singapore Sydney Tokyo

Senior Publishing Editor Jeremy Hayhurst
Editorial Coordinator Norah Donaghy
Production Services Manager Simon Crump
Senior Project Manager Julio Esperas
Composition, Design, Illustration, Indexing Frank H. Stephenson
Copyediting Simon & Associates
Cover Design Gary Ragaglia
Cover Image © William Whitehurst/CORBIS
Printing and Binding Courier
Cover Printer Phoenix Color

This book is printed on acid-free paper. ∞

Academic Press
An imprint of Elsevier Science
525 B Street, Suite 1900, San Diego, California 92101-4495, USA
http://www.academicpress.com

Academic Press
84 Theobald's Road, London WC1X 8RR, UK
http://www.academicpress.com

Academic Press
200 Wheeler Road, Burlington, Massachusetts 01803, USA
www.academicpressbooks.com

Library of Congress Control Number: 2003106049

International Standard Book Number: 0-12-665751-3

PRINTED IN THE UNITED STATES OF AMERICA
03 04 05 06 07 08 9 8 7 6 5 4 3 2 1

To my wife, Laurie, and our beautiful daughter, Myla.

Contents

Foreword

It was sometimes difficult for me to find a common ground with my high school chemistry teacher. We couldn't agree on what was appropriate attire or hair length for a sophomore. He had no respect for the notion that surfing is a viable career goal. He smoked way too much. But it was either in spite of him or because of him that I learned the mathematics involved in calculating the concentration of solutions, and I eventually agreed with him on the importance of numbers in the biological and chemical sciences. By the time I got out of that class, I was feeling as though I could have converted the speed of light into picomoles. As I undertook a career in molecular biology and teaching, it was what I learned in that class, more than in any other, that I have used on almost a daily basis.

Mathematics is a beautiful and elegant way of expressing order. I have heard it called a universal language. If true, then there are many dialects. There are any number of ways to approach a problem, no one of them necessarily more legitimate than another. I have always found interesting the passion and fervor people attach to their particular approach to mathematics. "Why do you do the problem **that** way?" I might be asked. "Clearly," my critic continues, "if you solve the problem **this** way, it's much quicker, more logical, and easier to follow. This is the only way that makes any sense." Well, maybe to them. Not everyone's brain works in the same way. In solving a problem in concentration, for example, it is probably amenable to solution by using a relationship of ratios, or $C_1V_1 = C_2V_2$, or the approach most often taken in this book. In actuality, they are all variations on a theme. Any one of them will get you the answer. Therein, I believe, lies the very beauty of mathematics.

It wasn't until this last year that I discovered the approach that I take to most of the problems encountered in the molecular biology laboratory has a name. It is called **dimensional analysis**. I always thought of it as "canceling terms." My brain is comfortable with this method. Many have tried to convert me to the use of the $C_1V_1 = C_2V_2$ approach, but all have failed. I have been chided and ridiculed by some for the manner in which I solve problems. I have been applauded by others. When I learned that the approach I take has the name **dimensional analysis**, I felt, in a way, like the person who visits the doctor with some inexplicable malady and who is reassured when the doctor attaches some Latin-sounding name to it. At least then, the individual knows that other people must also have the affliction, that it has been studied, and that there may even be a cure.

To present every possible way that each problem in this book could be solved would have made it too cumbersome. I make no apologies to those who might think that any one problem is better solved in another manner. To those of you who use this book as a companion in the laboratory, if you replace your values with the ones I have used in the example problems, you will do fine. You will have the numbers you need such that, for example, you will not be adding too little or too much salt that will send your cultured cells into osmotic shock or give you *Eco*RI star activity when digesting DNA.

My appreciation goes to my colleague Maria Abilock for bringing her critical eye to bear on this manuscript. She agreed to critique the draft in spite of the fact that she feels her way of solving problems is better than mine.

Frank H. Stephenson, Ph.D.

Scientific Notation and Metric Prefixes

1

Introduction

There are some 3,000,000,000 base pairs making up human genomic DNA within a haploid cell. If that DNA is isolated from such a cell, it will weigh approximately 0.0000000000035 grams. To amplify a specific segment of that purified DNA using the polymerase chain reaction (PCR), 0.00000000001 moles of each of two primers can be added to a reaction that can produce, following some 30 cycles of the PCR, over 1,000,000,000 copies of the target gene.

On a day-to-day basis, molecular biologists work with extremes of numbers far outside the experience of conventional life. To allow them to more easily cope with calculations involving extraordinary values, two shorthand methods have been adopted that bring both enormous and infinitesimal quantities back into the realm of manageability. These methods use scientific notation and metric prefixes. They require the use of exponents and an understanding of significant digits.

Significant Digits

Certain techniques in molecular biology, as in other disciplines of science, rely on types of instrumentation capable of providing precise measurements. An indication of the level of precision provided by any particular instrument is given by the number of digits expressed in its readout. The numerals of a measurement representing actual limits of precision are referred to as **significant digits**.

Although a zero can be as legitimate a value as the integers 1 through 9, significant digits are usually nonzero numerals. Without information on how a measurement was made or on the precision of the instrument used to make it, zeros to the left of the decimal point trailing one or more nonzero numerals are assumed not to be significant. For example, in stating that the human genome is 3,000,000,000 base pairs in length, the only significant digit in the number is the 3. The nine zeros are not significant. Likewise, zeros to the right of the decimal point preceding a set of nonzero numerals are assumed not to be significant. If we determine that the DNA within a sperm cell weighs 0.0000000000035 grams, only the 3 and the 5 are significant digits. The 11 zeros preceding these numerals are not significant.

Problem 1.1 How many significant digits are there in each of the following measurements?

a) 3,001,000,000 bp (base pairs)
b) 0.00304 grams
c) 0.000210 liters*
* Volume delivered with a calibrated micropipettor.

Solution 1.1

Given Number	Number of Significant Digits	The Significant Digits Are:
a) 3,001,000,000 bp	4	3001
b) 0.00304 grams	3	304
c) 0.000210 liters	3	210

Rounding Off Significant Digits in Calculations

When two or more measurements are used in a calculation, the result can only be as accurate as the least precise value. To accommodate this necessity, the number obtained as solution to a computation should be rounded off to reflect the weakest level of precision. The following guidelines will help determine the extent to which a numerical result should be rounded off.

Guidelines for Rounding Off Significant Digits

1. When adding or subtracting numbers, the result should be rounded off so that it has the same number of significant digits to the right of the decimal as that number used in the computation with the fewest significant digits to the right of the decimal.

2. When multiplying or dividing numbers, the result should be rounded off so that it contains only as many significant digits as that number in the calculation with the fewest significant digits.

Problem 1.2 Perform the following calculations, and express the answer using the guidelines for rounding off significant digits described in the preceding box.

a) 0.2884 g + 28.3 g
b) 3.4 cm × 8.115 cm
c) 1.2 L ÷ 0.155 L

Solution 1.2

a) 0.2884 g + 28.3 g = 28.5884 g

The sum is rounded off to show the same number of significant digits to the right of the decimal point as that number in the equation with the fewest significant digits to the right of the decimal point. (In this case, the value 28.3 has one significant digit to the right of the decimal point.)

28.5884 g is rounded off to 28.6 g.

b) 3.4 cm x 8.115 cm = 27.591 cm^2

The answer is rounded off to two significant digits since there are as few as two significant digits in one of the multiplied numbers (3.4 cm).

27.591 cm^2 is rounded off to 28 cm^2.

c) 1.2 L ÷ 0.155 L = 7.742 L

The quotient is rounded off to two significant digits since there are as few as two significant digits in one of the values (1.2 L) used in the equation.

7.742 L is rounded off to 7.7 L.

Exponents and Scientific Notation

An **exponent** is a number written above and to the right of (and smaller than) another number (called the **base**) to indicate the power to which the base is to be raised. Exponents of base 10 are used in scientific notation to express very large or vary small numbers in a shorthand form. For example, for the value 10^3, 10 is the base and 3 is the exponent. This means that 10 is multiplied by itself three times ($10^3 = 10 \times 10 \times 10 = 1000$). For numbers less than 1.0, a negative exponent is used to express values as a reciprocal of base 10. For example,

$$10^{-3} = \frac{1}{10^3} = \frac{1}{10 \times 10 \times 10} = \frac{1}{1000} = 0.001$$

Expressing Numbers in Scientific Notation

To express a number in scientific notation:

1. Move the decimal point to the right of the leftmost nonzero digit. Count the number of places the decimal has been moved from its original position.

2. Write the new number to include all numbers between the leftmost and rightmost significant (nonzero) figures. Drop all zeros lying outside these integers.

3. Place a multiplication sign and the number 10 to the right of the significant integers. Use an exponent to indicate the number of places the decimal point has been moved.
   a. For numbers greater than 10 (where the decimal was moved to the left), use a positive exponent.
   b. For numbers less than 1 (where the decimal was moved to the right), use a negative exponent.

Problem 1.3 Write the following numbers in scientific notation.

a) 3,001,000,000
b) 78
c) 60.23×10^{22}

Solution 1.3

a) 3,001,000,000

Move the decimal to the left nine places so that it is positioned to the right of the leftmost nonzero digit.

3.001000000

Write the new number to include all nonzero significant figures, and drop all zeros outside of these numerals. Multiply the new number by 10, and use a positive 9 as the exponent since the given number is greater than 10 and the decimal was moved to the left nine positions.

$$3{,}001{,}000{,}000 = 3.001 \times 10^9$$

b) 78

Move the decimal to the left one place so that it is positioned to the right of the leftmost nonzero digit. Multiply the new number by 10, and use a positive 1 as an exponent since the given number is greater than 10 and the decimal was moved to the left one position.

$$78 = 7.8 \times 10^1$$

c) 60.23×10^{22}

Move the decimal to the left one place so that it is positioned to the right of the leftmost nonzero digit. Since the decimal was moved one position to the left, add 1 to the exponent (22 + 1 = 23 = new exponent value).

$$60.23 \times 10^{22} = 6.023 \times 10^{23}$$

Problem 1.4 Write the following numbers in scientific notation.

a) 0.000000000015
b) 0.0000500042
c) 437.28×10^{-7}

Solution 1.4

a) 0.000000000015

Move the decimal to the right 11 places so that it is positioned to the right of the leftmost nonzero digit. Write the new number to include all numbers between the leftmost and rightmost significant (nonzero) figures. Drop all zeros lying outside these numerals. Multiply the number by 10, and use a negative 11 as the exponent since the original number is less than 1 and the decimal was moved to the right by 11 places.

$$0.000000000015 = 1.5 \times 10^{-11}$$

b) 0.0000500042

Move the decimal to the right five positions so that it is positioned to the right of the leftmost nonzero digit. Drop all zeros lying outside the leftmost and rightmost nonzero digits. Multiply the number by 10 and use a negative 5 exponent since the original number is less than 1 and the decimal point was moved to the right five positions.

$$0.0000500042 = 5.00042 \times 10^{-5}$$

c) 437.28×10^{-7}

Move the decimal point two places to the left so that it is positioned to the right of the leftmost nonzero digit. Since the decimal is moved two places to the left, add a positive 2 to the exponent value ($-7 + 2 = -5$).

$$437.28 \times 10^{-7} = 4.3728 \times 10^{-5}$$

Converting Numbers from Scientific Notation to Decimal Notation

To change a number expressed in scientific notation to decimal form:

1. If the exponent of 10 is positive, move the decimal point to the right the same number of positions as the value of the exponent. If necessary, add zeros to the right of the significant digits to hold positions from the decimal point.

2. If the exponent of 10 is negative, move the decimal point to the left the same number of positions as the value of the exponent. If necessary, add zeros to the left of the significant digits to hold positions from the decimal point.

Problem 1.5 Write the following numbers in decimal form.

a) 4.37×10^5
b) 2×10^1
c) 23.4×10^7
d) 3.2×10^{-4}

Solution 1.5

a) 4.37×10^5

Move the decimal point five places to the right, adding three zeros to hold the decimal's place from its former position.

$$4.37 \times 10^5 = 437{,}000.0$$

b) 2×10^1

Move the decimal point one position to the right, adding one zero to the right of the significant digit to hold the decimal point's new position.

$$2 \times 10^1 = 20.0$$

c) 23.4×10^7

Move the decimal point seven places to the right, adding six zeros to hold the decimal point's position.

$$23.4 \times 10^7 = 234{,}000{,}000.0$$

d) 3.2×10^{-4}

The decimal point is moved four places to the left. Zeros are added to hold the decimal point's position.

$$3.2 \times 10^{-4} = 0.00032$$

Adding and Subtracting Numbers Written in Scientific Notation

When adding or subtracting numbers expressed in scientific notation, it is simplest first to convert the numbers in the equation to the same power of 10 as that of the highest exponent. The exponent value then does not change when the computation is finally performed.

Problem 1.6 Perform the following computations.

a) $(8 \times 10^4) + (7 \times 10^4)$
b) $(2 \times 10^3) + (3 \times 10^1)$
c) $(6 \times 10^{-2}) + (8 \times 10^{-3})$
d) $(3.9 \times 10^{-4}) - (3.7 \times 10^{-4})$
e) $(2.4 \times 10^{-3}) - (1.1 \times 10^{-4})$

Solution 1.6

a) $(8 \times 10^4) + (7 \times 10^4)$

$= 15 \times 10^4$ — Numbers added.
$= 1.5 \times 10^5$ — Number rewritten in standard scientific notation form.
$= 2 \times 10^5$ — Number rounded off to 1 significant digit.

On the calculator:

8	EXP	4	+	7	EXP	4
=						

b) $(2 \times 10^3) + (3 \times 10^1)$

$= (2 \times 10^3) + (0.03 \times 10^3)$ — Number with lowest exponent value expressed in terms of that of the largest exponent value.
$= 2.03 \times 10^3$ — Numbers are added.
$= 2 \times 10^3$ — Number rounded off to 1 significant digit.

c) $(6 \times 10^{-2}) + (8 \times 10^{-3})$

$= (6 \times 10^{-2}) + (0.8 \times 10^{-2})$	Exponents converted to the same values.
$= 6.8 \times 10^{-2}$	Numbers are added.
$= 7 \times 10^{-2}$	Number rounded off to 1 significant digit.

On the calculator:

[6] [EXP] [2] [+/−] [+] [8] [EXP]

[3] [+/−] [=]

d) $(3.9 \times 10^{-4}) - (3.7 \times 10^{-4})$

$= 0.2 \times 10^{-4}$	Numbers are subtracted.
$= 2 \times 10^{-5}$	Number rewritten in standard scientific notation form.

e) $(2.4 \times 10^{-3}) - (1.1 \times 10^{-4})$

$= (2.4 \times 10^{-3}) - (0.11 \times 10^{-3})$	Exponents converted to the same values.
$= 2.29 \times 10^{-3}$	Numbers are subtracted.
$= 2.3 \times 10^{-3}$	Number rounded off to show only one significant digit to the right of the decimal point.

Multiplying and Dividing Numbers Written in Scientific Notation

Exponent laws used in multiplication and division for numbers written in scientific notation include:

The Product Rule: When multiplying using scientific notation, the exponents are added.

The Quotient Rule: When dividing using scientific notation, the exponent of the denominator is subtracted from the exponent of the numerator.

When working with the next set of problems, the following laws of mathematics will be helpful:

The Commutative Law for Multiplication: The result of a multiplication is not dependent on the order in which the numbers are multiplied. For example,

$$3 \times 2 = 2 \times 3$$

The Associative Law for Multiplication: The result of a multiplication is not dependent on how the numbers are grouped. For example,

$$3 \times (2 \times 4) = (3 \times 2) \times 4$$

Problem 1.7 Calculate the product.

a) $(3 \times 10^4) \times (5 \times 10^2)$
b) $(2 \times 10^3) \times (6 \times 10^{-5})$
c) $(4 \times 10^{-2}) \times (2 \times 10^{-3})$

Solution 1.7

a) $(3 \times 10^4) \times (5 \times 10^2)$

$= (3 \times 5) \times (10^4 \times 10^2)$	Use Commutative and Associative laws to group like terms.
$= 15 \times 10^6$	Exponents are added.
$= 1.5 \times 10^7$	Number written in standard scientific notation form.
$= 2 \times 10^7$	Number rounded off to one significant digit.

On the calculator:

[3] [EXP] [4] [X] [5] [EXP] [2]

[=]

b) $(2 \times 10^3) \times (6 \times 10^{-5})$

$= (2 \times 6) \times (10^3 \times 10^{-5})$	Use Commutative and Associative laws to group like terms.
$= 12 \times 10^{-2}$	Exponents are added.

$= 1.2 \times 10^{-1}$	Number written in standard scientific notation form.
$= 1 \times 10^{-1}$	Number rounded off to one significant digit.

On the calculator:

[2] [EXP] [3] [X] [6] [EXP] [5]

[+/−] [=]

c) $(4 \times 10^{-2}) \times (2 \times 10^{-3})$

$= (4 \times 2) \times (10^{-2} \times 10^{-3})$	Use Commutative and Associative laws to group like terms.
$= 8 \times 10^{-2+(-3)}$	
$= 8 \times 10^{-5}$	Exponents are added.

On the calculator:

[4] [EXP] [2] [+/−] [X] [2] [EXP]

[3] [+/−] [=]

Problem 1.8 Find the quotient.

a) $\frac{8 \times 10^{4}}{2 \times 10^{2}}$

b) $\frac{5 \times 10^{8}}{3 \times 10^{-4}}$

c) $\frac{8.2 \times 10^{-6}}{3.6 \times 10^{4}}$

d) $\frac{9 \times 10^{-5}}{2.5 \times 10^{-3}}$

Solution 1.8

a) $\dfrac{8\times10^{4}}{2\times10^{2}}$

The exponent of the denominator is subtracted from the exponent of the numerator.

$= \dfrac{8}{2}\times10^{4-2}$

$= 4\times10^{2}$

On the calculator:

b) $\dfrac{5\times10^{8}}{3\times10^{-4}}$

The exponent of the denominator is subtracted from the exponent of the numerator.

$= \dfrac{5}{3}\times 10^{8-(-4)}$

$= 1.67 \times 10^{8-(-4)}$ Exponents: $8-(-4) = 8+4 = 12$.

$= 2 \times 10^{12}$ Number rounded off to one significant digit.

On the calculator:

5	EXP	8	÷	3	EXP	4
+/−	=					

c) $\dfrac{8.2\times10^{-6}}{3.6\times10^{4}}$

The exponent of the denominator is subtracted from the exponent of the numerator.

$= \dfrac{8.2}{3.6}\times 10^{-6-(+4)}$

$= 2.3 \times 10^{-10}$ Number rounded off to two significant digits.

Exponent: $-6-(+4) = -6+(-4) = -10$.

On the calculator:

8	•	2	EXP	6	+/−	÷
3	•	6	EXP	4	=	

d) $\dfrac{9 \times 10^{-5}}{2.5 \times 10^{-3}}$

The exponent of the denominator is subtracted from the exponent of the numerator.

$= \dfrac{9}{2.5} \times 10^{-5-(-3)}$

$= 3.6 \times 10^{-2}$

$= 4 \times 10^{-2}$ Number rounded off to one significant digit.

On the calculator:

9	EXP	5	+/−	÷	2	•
5	EXP	3	+/−	=		

Metric Prefixes

A **metric prefix** is a shorthand notation used to denote very large or vary small values of a basic unit as an alternative to expressing them as powers of 10. Basic units frequently used in the biological sciences include meters, grams, moles, and liters. Because of their simplicity, metric prefixes have found wide application in molecular biology. The following table lists the most frequently used prefixes and the values they represent.

Metric Prefix	Abbreviation	Power of 10
giga-	G	10^{9}
mega-	M	10^{6}
kilo-	k	10^{3}
milli-	m	10^{-3}
micro-	μ	10^{-6}
nano-	n	10^{-9}
pico-	p	10^{-12}
femto-	f	10^{-15}
atto-	a	10^{-18}

As shown in the table, 1 nanogram (ng) is equivalent to 1×10^{-9} grams. There are, therefore, 1×10^{9} nanograms per gram (the reciprocal of 1×10^{-9}; $1/1 \times 10^{9} = 1 \times 10^{-9}$). Likewise, since one microliter (μL) is equivalent to 1×10^{-6} liters, there are 1×10^{6} μL per liter.

When expressing quantities with metric prefixes, the prefix is usually chosen so that the value can be written as a number greater than 1.0 but less than 1000. For example, it is conventional to express 0.00000005 grams as 50 ng rather than 0.05 μg or 50,000 pg.

Conversion Factors and Canceling Terms

Translating a measurement expressed with one metric prefix into an equivalent value expressed using a different metric prefix is called a **conversion**. These are performed mathematically by using a conversion factor relating the two different terms. A conversion factor is a numerical ratio equal to 1. For example,

$$\frac{1 \times 10^{6}\ \mu g}{g} \quad \text{and} \quad \frac{1\ g}{1 \times 10^{6}\ \mu g}$$

are conversion factors, both equal to 1. They can be used to convert grams to micrograms or micrograms to grams, respectively. The final metric prefix expression desired should appear in the equation as a numerator value in the conversion factor. Since multiplication or division by the number 1 does not change the value of the original quantity, any quantity can be either multiplied or divided by a conversion factor and the result will still be equal to the original quantity; only the metric prefix will be changed.

When performing conversions between values expressed with different metric prefixes, the calculations can be simplified when factors of 1 or identical units are canceled. A factor of 1 is any expression in which a term is divided by itself. For example, $1 \times 10^{6}/1 \times 10^{6}$ is a factor of 1. Likewise, 1 liter/1 liter is a factor of 1. If, in a conversion, identical terms appear anywhere in the equation on one side of the equals sign as both a numerator and a denominator, they can be canceled. For example, if converting 5×10^{-4} liters to microliters, an equation can be set up so that identical terms (in this case, liters) can be canceled to leave μL as a numerator value.

$5 \times 10^{-4}\,\text{L} = n\ \mu\text{L}$ — Solve for n.

Use the conversion factor relating liters and microliters with microliters as a numerator value. Identical terms in a numerator and a denominator are canceled. (Remember, 5×10^{-4} liters is the same as 5×10^{-4} L/1. 5×10^{-4} liters, therefore, is a numerator.)

$$5\times10^{-4}\ \text{L}\times\frac{1\times10^{6}\ \mu\text{L}}{\text{L}} = n\ \mu\text{L}$$

$$(5\times1)(10^{-4}\times10^{6})\ \mu\text{L} = n\ \mu\text{L}$$

Group like terms.

$$5\times10^{-4+6}\ \mu\text{L} = n\ \mu\text{L}$$

Numerator values are multiplied.

$$5\times10^{2}\ \mu\text{L} = n\ \mu\text{L}$$

Therefore, 5×10^{-4} liters is equivalent to 5×10^{2} µL.

Problem 1.9

a) There are approximately 6×10^{9} base pairs (bp) per human diploid genome. What is this number expressed as kilobase pairs (kb)?

b) Convert 0.03 µg into nanograms.

c) Convert 0.0025 mL into µL.

Solution 1.9

a) 6×10^{9} bp $= n$ kb — Solve for n.

Multiply by a conversion factor relating kb to bp with kb as a numerator:

$$6\times10^{9}\ \text{bp}\times\frac{1\ \text{kb}}{1\times10^{3}\ \text{bp}} = n\ \text{kb}$$

Cancel identical terms (bp) appearing as numerator and denominator, leaving kb as a numerator value.

$$\frac{(6\times10^{9})(1 \text{ kb})}{1\times10^{3}} = n \text{ kb}$$

The exponent of the denominator is subtracted from the exponent of the numerator.

$$\frac{6}{1}\times10^{9-3} \text{ kb} = 6\times10^{6} \text{ kb} = n \text{ kb}$$

Therefore, 6×10^{9} base pairs is equivalent to 6×10^{6} kb.

b) 0.03 µg = n ng Solve for n.

Multiply by conversion factors relating g to µg and ng to g with ng as a numerator. Convert 0.03 µg to its equivalent in scientific notation (3×10^{-2} µg):

$$3\times10^{-2} \text{ µg}\times\frac{1 \text{ g}}{1\times10^{6} \text{ µg}}\times\frac{1\times10^{9} \text{ ng}}{\text{g}} = n \text{ ng}$$

Cancel identical terms appearing as numerator and denominator, leaving ng as a numerator value; multiply numerator and denominator values; and then group like terms.

$$\frac{(3\times1\times1)(10^{-2}\times10^{9}) \text{ ng}}{(1\times1)(10^{6})} = n \text{ ng}$$

Numerator exponents are added.

$$\frac{3\times10^{-2+9} \text{ ng}}{1\times10^{6}} = \frac{3\times10^{7} \text{ ng}}{1\times10^{6}} = n \text{ ng}$$

The denominator exponent is subtracted from the numerator exponent.

$$\frac{3}{1}\times10^{7-6} \text{ ng} = 3\times10^{1} \text{ ng} = n \text{ ng}$$

Therefore, 0.03 µg is equivalent to 30 (3×10^{1}) ng.

c) 0.0025 mL = n µL Solve for n.

Convert 0.0025 mL into scientific notation. Multiply by conversion factors relating L to mL and µL to L with µL as a numerator.

$$2.5 \times 10^{-3} \text{ mL} \times \frac{1 \text{ L}}{1 \times 10^{3} \text{ mL}} \times \frac{1 \times 10^{6} \ \mu\text{L}}{1 \text{ L}} = n \ \mu\text{L}$$

Cancel identical terms appearing as numerator and denominator, leaving µL as a numerator value. Multiply numerator values and denominator values. Group like terms.

$$\frac{(2.5 \times 1 \times 1)(10^{-3} \times 10^{6}) \ \mu\text{L}}{(1 \times 1)(10^{3})} = n \ \mu\text{L}$$

Numerator exponents are added.

$$\frac{2.5 \times 10^{-3+6} \ \mu\text{L}}{1 \times 10^{3}} = \frac{2.5 \times 10^{3} \ \mu\text{L}}{1 \times 10^{3}} = n \ \mu\text{L}$$

The denominator exponent is subtracted from the numerator exponent.

$$\frac{2.5}{1} \times 10^{3-3} \ \mu\text{L} = 2.5 \times 10^{0} \ \mu\text{L} = 2.5 \ \mu\text{L} = n \ \mu\text{L}$$

Therefore, 0.0025 mL is equivalent to 2.5 µL.

Solutions, Mixtures, and Media

2

Introduction

Whether it is an organism or an enzyme, most biological activities function at their optimum only within a narrow range of environmental conditions. From growing cells in culture to sequencing of a cloned DNA fragment or assaying an enzyme's activity, the success or failure of an experiment can hinge on paying careful attention to a reaction's components. This section outlines the mathematics involved in making solutions.

Calculating Dilutions: A General Approach

Concentration is defined as an amount of some substance per a set volume:

$$\text{concentration} = \frac{\text{amount}}{\text{volume}}$$

Most laboratories have found it convenient to prepare concentrated stock solutions of commonly used reagents, those found as components in a large variety of buffers or reaction mixes. Such stock solutions may include 1 *M* Tris, pH 8.0, 500 m*M* EDTA, 20% sodium dodecylsulfate (SDS), 1 *M* $MgCl_2$, and any number of others. A specific volume of a stock solution at a particular concentration can be added to a buffer or reagent mixture so that it contains that component at some concentration less than that in the stock. For example, a stock solution of 95% ethanol can be used to prepare a solution of 70% ethanol. Since a higher percent solution (more concentrated) is being used to prepare a lower percent (less concentrated) solution, a **dilution** of the stock solution is being performed.

There are several methods that can be used to calculate the concentration of a diluted reagent. No one approach is necessarily more valid than another. Typically, the method chosen by an individual has more to do with how his or her brain approaches mathematical problems than with the legitimacy of the procedure. One approach is to use the equation $C_1V_1 = C_2V_2$ where

C_1 is the initial concentration of the stock solution,
V_1 is the amount of stock solution taken to perform the dilution,
C_2 is the concentration of the diluted sample, and
V_2 is the final, total volume of the diluted sample.

For example, if you were asked how many μL of 20% sugar should be used to make 2 mL of 5% sucrose, the $C_1V_1 = C_2V_2$ equation could be used. However, to use this approach, all units must be the same. Therefore, you first need to convert 2 mL into a microliter amount. This can be done as follows:

$$2\ \text{mL} \times \frac{1000\ \mu\text{L}}{1\ \text{mL}} = 2000\ \ \mu\text{L}$$

C_1, then, is equal to 20%, V_1 is the volume you wish to calculate, C_2 is 5%, and V_2 is 2000 μL. The calculation is then performed as follows:

$$C_1V_1 = C_2V_2$$
$$(20\%)V_1 = (5\%)(2000\ \ \text{mL})$$

Solving for V_1 gives the following result.

$$V_1 = \frac{(5\%)(2000\ \ \mu\text{L})}{20\%} = 500\ \ \mu\text{L}$$

The % units cancel since they are in both the numerator and the denominator of the equation, leaving μL as the remaining unit. Therefore, you would need 500 μL of 20% sucrose plus 1500 μL (2000 μL – 500 μL = 1500 μL) of water to make a 5% sucrose solution from a 20% sucrose solution.

Dimensional analysis is another general approach to solving problems of concentration. In this method, an equation is set up such that the known concentration of the stock and all volume relationships appear on the left side of the equation and the final desired concentration is placed on the right side. Conversion factors are actually part of the equation. Terms are set up as numerator or denominator values such that all terms cancel except for that describing concentration. A dimensional analysis equation is set up in the following manner.

$$\text{starting concentration} \times \text{conversion factor} \times \frac{\text{unknown volume}}{\text{final volume}} = \text{desired concentration}$$

Using the dimensional analysis approach, the problem of discovering how many microliters of 20% sucrose are needed to make 2 mL of 5% sucrose is written as follows:

$$20\% \times \frac{1\ \text{mL}}{1000\ \mu\text{L}} \times \frac{x\ \mu\text{L}}{2\ \text{mL}} = 5\%$$

Notice that all terms on the left side of the equation will cancel except for %. Solving for x μL gives the following result:

$$\frac{(20\%)x}{2000} = 5\%$$

$$x = \frac{(5\%)(2000)}{20\%} = 500$$

Since x is a μL amount, you need 500 μL of 20% sucrose in a final volume of 2 mL to make 5% sucrose. Notice how similar the last step of the solution to this equation is to the last step of the equation using the $C_1V_1 = C_2V_2$ approach.

Making a conversion factor part of the equation obviates the need for performing two separate calculations, as is required when using the $C_1V_1 = C_2V_2$ approach. For this reason, dimensional analysis is the method used for solving problems of concentration throughout this book.

Concentrations by a Factor of X

The concentration of a solution can be expressed as a multiple of its standard working concentration. For example, many buffers used for agarose or acrylamide gel electrophoresis are prepared as solutions 10-fold (10X) more concentrated than their standard running concentration (1X). In a 10X buffer, each component of that buffer is 10-fold more concentrated than in the 1X solution. To prepare a 1X working buffer, a dilution of the more concentrated 10X stock is performed in water to achieve the desired volume. To prepare 1000 mL (1 L) of 1X Tris-borate-EDTA (TBE) gel running buffer from a 10X TBE concentrate, for example, add 100 mL of 10X solution to 900 mL of distilled water. This can be calculated as follows:

$$10\text{X Buffer} \times \frac{n \text{ mL}}{1000 \text{ mL}} = 1\text{X Buffer}$$

n mL of 10X buffer is diluted into a total volume of 1000 mL to give a final concentration of 1X. Solve for n.

$$\frac{10\text{X}n}{1000} = 1\text{X}$$

Multiply numerator values.

$$(1000) \times \frac{10\text{X}n}{1000} = 1\text{X } (1000)$$

Use the Multiplication Property of Equality (see the following box) to multiply each side of the equation by 1000. This cancels out the 1000 in the denominator on the left side of the equals sign.

$$10\text{X}n = 1000\text{X}$$

$$\frac{10\text{X}n}{10\text{X}} = \frac{1000\text{X}}{10\text{X}}$$

Divide each side of the equation by 10X. (Again, this uses the Multiplication Property of Equality.)

$$n = 100$$

The X terms cancel since they appear in both the numerator and the denominator. This leaves n equal to 100.

Therefore, to make1000 mL of 1X buffer, add 100 mL of 10X buffer stock to 900 mL of distilled water (1000 mL – 100 mL contributed by the 10X buffer stock = 900 mL).

Multiplication Property of Equality

Both sides of an equation may be multiplied by the same nonzero quantity to produce equivalent equations. This property also applies to division: both sides of an equation can be divided by the same nonzero quantity to produce equivalent equations.

Problem 2.1 How are 640 mL of 0.5X buffer prepared from an 8X stock?

Solution 2.1 We start with a stock of 8X buffer. We want to know how many milliliters of the 8X buffer should be in a final volume of 640 mL to give us a buffer having a concentration of 0.5X. This relationship can be expressed mathematically as follows:

$$8\text{X buffer} \times \frac{n \text{ mL}}{640 \text{ mL}} = 0.5\text{X buffer}$$

Solve for n.

$$\frac{8\text{X}n}{640} = 0.5\text{X}$$

Multiply numerator values on the left side of the equation. Since the mL terms appear in both the numerator and the denominator, they cancel out.

$8Xn = 320X$	Multiply each side of the equation by 640.
$n = \frac{320X}{8X} = 40$	Divide each side of the equation by 8X. The X terms, since they appear in both the numerator and the denominator, cancel.

Therefore, add 40 mL of 8X stock to 600 mL of distilled water to prepare a total of 640 mL of 0.5X buffer (640 mL final volume – 40 mL 8X stock = 600 mL volume to be taken by water).

Preparing Percent Solutions

Many reagents are prepared as a percent of solute (such as salt, cesium chloride, or sodium hydroxide) dissolved in solution. Percent, by definition, means "per 100." 12%, therefore, means 12 per 100, or 12 out of every 100. 12% may also be written as the decimal 0.12 (derived from the fraction 12/100 = 0.12).

Depending on the solute's initial physical state, its concentration can be expressed as a weight per volume percent (% w/v) or a volume per volume percent (% v/v). A percentage in weight per volume refers to the weight of solute (in grams) in a total of 100 mL of solution. A percentage in volume per volume refers to the amount of liquid solute (in mL) in a final volume of 100 mL of solution.

Most microbiology laboratories will stock a solution of 20% (w/v) glucose for use as a carbon source in bacterial growth media. To prepare 100 mL of 20% (w/v) glucose, 20 grams of glucose are dissolved in enough distilled water so that the final volume of the solution, with the glucose completely dissolved, is 100 mL.

Problem 2.2 How can the following solutions be prepared?

a) 100 mL of 40% (w/v) polyethylene glycol (PEG) 8000
b) 47 mL of a 7% (w/v) solution of sodium chloride
c) 200 mL of a 95% (v/v) solution of ethanol

Solution 2.2

a) Weigh out 40 grams of PEG 8000 and dissolve in distilled water so that the final volume of the solution, with the PEG 8000 completely dissolved, is 100 mL. This is most conveniently done by initially dissolving the PEG 8000 in approximately 60 mL of distilled water. When the granules are dissolved, pour the solution into a 100-mL graduated cylinder and bring the volume up to the 100-mL mark with distilled water.

b) First, 7% of 47 must be calculated. This is done by multiplying 47 by 0.07 (the decimal form of 7%; $\frac{7}{100} = 0.07$):

$$0.07 \times 47 = 3.29$$

Therefore, to prepare 47 mL of 7% sodium chloride, weigh out 3.29 grams of sodium chloride and dissolve the crystals in some volume of distilled water less than 47 mL, a volume measured so that, when the 3.29 grams of NaCl are added, it does not exceed 47 mL. When the sodium chloride is completely dissolved, dispense the solution into a 50-mL graduated cylinder and bring the final volume up to 47 mL with distilled water.

c) 95% of 200 mL is calculated by multiplying 0.95 (the decimal form of 95%) by 200:

$$0.95 \times 200 = 190$$

Therefore, to prepare 200 mL of 95% ethanol, measure 190 mL of 100% (200 proof) ethanol and add 10 mL of distilled water to bring the final volume to 200 mL.

Diluting Percent Solutions

When approaching a dilution problem involving percentages, express the percent solutions as fractions of 100. The problem can be written as an equation in which the concentration of the stock solution ("what you have") is positioned on the left side of the equation and the desired final concentration ("what you want") is on the right side of the equation. The unknown volume (x) of the stock solution to add to the volume of the final mixture should also be expressed as a fraction (with x as a numerator and the final desired volume as a denominator). This part of the equation should also be positioned on the left side of the equals sign. For example, if 30 mL of 70% ethanol is to be prepared from a 95% ethanol stock solution, the following equation can be written:

$$\frac{95}{100} \times \frac{x \text{ mL}}{30 \text{ mL}} = \frac{70}{100}$$

You then solve for x.

$$\frac{95x}{3000} = \frac{70}{100}$$

Multiply numerators together and multiply denominators together. The mL terms, since they are present in both the numerator and the denominator, cancel.

$$\frac{3000}{1} \times \frac{95x}{3000} = \frac{70}{100} \times \frac{3000}{1}$$

Multiply both sides of the equation by 3000.

$$95x = \frac{210,000}{100}$$

Simplify the equation.

$$95x = 2100$$

$$\frac{95x}{95} = \frac{2100}{95}$$

Divide each side of the equation by 95.

$$x = 22$$

Round off to two significant figures.

Therefore, to prepare 30 mL of 70% ethanol using a 95% ethanol stock solution, combine 22 mL of 95% ethanol stock with 8 mL of distilled water.

Problem 2.3 If 25 grams of NaCl are dissolved into a final volume of 500 mL, what is the percent (w/v) concentration of NaCl in the solution?

Solution 2.3 The concentration of NaCl is 25 g/500 mL (w/v). To determine the % (w/v) of the solution, we need to know how many grams of NaCl are in 100 mL. We can set up an equation of two ratios in which x represents the unknown number of grams. This relationship is read "x grams is to 100 mL as 25 grams is to 500 mL":

$$\frac{x \text{ g}}{100 \text{ mL}} = \frac{25 \text{ g}}{500 \text{ mL}}$$

Solving for x gives the following result:

$$x \text{ g} = \frac{(25 \text{ g})(100 \text{ mL})}{500 \text{ mL}} = \frac{2500 \text{ g}}{500} = 5 \text{ g}$$

Therefore, there are 5 grams of NaCl in 100 mL (5 g/100 mL), which is equivalent to a 5% solution.

Problem 2.4 If 8 mL of distilled water are added to 2 mL of 95% ethanol, what is the concentration of the diluted ethanol solution?

Solution 2.4 The total volume of the solution is 8 mL + 2 mL = 10 mL. This volume should appear as a denominator on the left side of the equation. This dilution is the same as if 2 mL of 95% ethanol were added to 8 mL of water. Either way, it is a quantity of the 95% ethanol stock that is used to make the dilution. The "2 mL," therefore, should appear as the numerator in the volume expression on the left side of the equation:

$$\frac{95}{100} \times \frac{2 \text{ mL}}{10 \text{ mL}} = \frac{x}{100}$$

$$\frac{190}{1000} = \frac{x}{100}$$

The mL terms cancel. Multiply numerator values and denominator values on the left side of the equation.

$$0.19 = \frac{x}{100}$$

Simplify the equation.

$$19 = x$$

Multiply both sides of the equation by 100.

If x in the original equation is replaced by19, it is seen that the new concentration of ethanol in this diluted sample is 19/100, or 19%.

Problem 2.5 How many microliters of 20% SDS are required to bring 1.5 mL of solution to 0.5%?

Solution 2.5 In previous examples, there was control over how much water we could add in preparing the dilution to bring the sample to the desired concentration. In this example, however, a fixed volume (1.5 mL) is used as a starting sample and must be brought to the desired concentration. Solving this problem will require the use of the Addition Property of Equality (see the following box).

Addition Property of Equality

You may add (or subtract) the same quantity to (from) both sides of an equation to produce equivalent equations. For any real numbers a, b, and c, if $a = b$, then $a + c = b + c$, and $a - c = b - c$.

Since concentration, by definition, is the amount of a particular component in a specified volume, by adding a quantity of a stock solution to a fixed volume, the final volume is changed by that amount and the concentration is changed accordingly. The amount of stock solution (x mL) added in the process of the dilution must also be figured into the final volume, as follows.

$$\frac{20}{100} \times \frac{x \text{ mL}}{1.5 \text{ mL} + x \text{ mL}} = \frac{0.5}{100}$$

$$\frac{20x}{150 + 100x} = \frac{0.5}{100}$$

Multiply numerators and denominators. The mL terms cancel out.

$$\frac{150 + 100x}{1} \times \frac{20x}{150 + 100x} = \frac{0.5}{100} \times \frac{150 + 100x}{1}$$

Multiply both sides of the equation by 150 + 100x.

$$20x = \frac{75 + 50x}{100}$$

Simplify the equation.

$$\frac{100}{1} \times 20x = \frac{75 + 50x}{100} \times \frac{100}{1}$$

Multiply both sides of the equation by 100.

$$2000x = 75 + 50x$$

Simplify.

$$1950x = 75$$

Subtract 50x from both sides of the equation (Addition Property of Equality).

$$x = \frac{75}{1950} = 0.03846 \text{ mL}$$

Divide both sides of the equation by 1950.

$$0.03846 \text{ mL} \times \frac{1000 \ \mu\text{L}}{\text{mL}} = 38.5 \ \mu\text{L}$$

Convert mL to µL and round off to one significant figure to the right of the decimal point.

Therefore, if 38.5 µL of 20% SDS are added to 1.5 mL, the SDS concentration of that sample will be 0.5% in a final volume of 1.5385 mL. If there were some other component in that initial 1.5 mL, the concentration of that component would change by the addition of the SDS. For example, if NaCl were present at a concentration of 0.2%, its concentration would be altered by the addition of more liquid. The initial solution of 1.5 mL would contain the following amount of sodium chloride:

$$\frac{0.2\ \text{g}}{100\ \text{mL}} \times 1.5\ \text{mL} = 0.003\ \text{g}$$

Therefore, 1.5 mL of 0.2% NaCl contains 0.003 grams of NaCl.

In a volume of 1.5385 mL (the volume after the SDS solution has been added), 0.003 grams of NaCl is equivalent to a 0.195% NaCl solution, as shown here:

$$\frac{0.003}{1.5385} \times 100 = 0.195\%$$

Moles and Molecular Weight: Definitions

A **mole** is equivalent to 6.023×10^{23} molecules. That molecule may be a pure elemental atom or a molecule consisting of a bound collection of atoms. For example, a mole of hydrogen is equivalent to 6.023×10^{23} molecules of hydrogen. A mole of glucose ($C_6H_{12}O_6$) is equivalent to 6.023×10^{23} molecules of glucose. The value 6.023×10^{23} is also known as **Avogadro's number**.

The **molecular weight** (**MW**, or **gram molecular weight**) of a substance is equivalent to the sum of its atomic weights. For example, the gram molecular weight of sodium chloride (NaCl) is 58.44, the atomic weight of Na (22.99 g) plus the atomic weight of chlorine (35.45 g). Atomic weights can be found in the periodic table of the elements. The molecular weight of a compound, as obtained commercially, is usually provided by the manufacturer and is printed on the container's label. On many reagent labels, a **formula weight** (**FW**) is given. For almost all applications in molecular biology, this value is used interchangeably with molecular weight.

Problem 2.6 What is the molecular weight of sodium hydroxide (NaOH)?

Solution 2.6

Atomic weight of Na	22.99
Atomic weight of O	16.00
Atomic weight of H	+ 1.01
Molecular weight of NaOH	40.00

Problem 2.7 What is the molecular weight of glucose ($C_6H_{12}O_6$)?

Solution 2.7 The atomic weight of each element in this compound must be multiplied by the number of times it is represented in the molecule:

Atomic weight of C = 12.01	
$12.01 \times 6 =$	72.06
Atomic weight of H = 1.01	
$1.01 \times 12 =$	12.12
Atomic weight of O = 16.00	
$16 \times 6 =$	+ 96.00
Molecular weight of $C_6H_{12}O_6$	180.18

Therefore, the molecular weight of glucose is 180.18.

Molarity

A 1 molar (1 *M*) solution contains the molecular weight of a substance (in grams) in 1 liter of solution. For example, the molecular weight of NaCl is 58.44. A 1 *M* solution of sodium chloride (NaCl), therefore, contains 58.44 grams of NaCl dissolved in a final volume of 1000 mL (1 L) water. A 2 *M* solution of NaCl contains twice that amount (116.88 grams) of NaCl dissolved in a final volume of 1000 mL water.

Problem 2.8 How are 200 mL of 0.3 *M* NaCl prepared?

Solution 2.8 The molecular weight of NaCl is 58.44. The first step in solving this problem is to calculate how many grams are needed for 1 L of a 0.3 *M* solution. This can be done by setting up a ratio stating, "58.44 grams is to 1 *M* as *x* grams is to 0.3 *M*." This relationship, expressed mathematically, can be written as follows. We then solve for *x*.

$$\frac{58.44 \text{ g}}{1\ M} = \frac{x \text{ g}}{0.3\ M}$$

Because units on both sides of the equation are equivalent (if we were to multiply one side by the other, all terms would cancel), we will disregard them. Multiplying both sides of the equation by 0.3 gives

$$\frac{0.3}{1} \times \frac{58.44}{1} = \frac{x}{0.3} \times \frac{0.3}{1}$$

$$17.53 = x$$

Therefore, to prepare 1 L of 0.3 *M* NaCl, 17.53 grams of NaCl are required.

Another ratio can now be written to calculate how many grams of NaCl are needed if 200 mL of a 0.3 *M* NaCl solution are being prepared. It can be expressed verbally as "17.53 grams is to1000 mL as *x* grams is to 200 mL," or written in mathematical terms:

$$\frac{17.53 \text{ g}}{1000 \text{ mL}} = \frac{x \text{ g}}{200 \text{ mL}}$$

$$\frac{17.53 \times 200}{1000} = x$$

Multiply both sides of the equation by 200 to cancel out the denominator on the right side of the equation and to isolate *x*.

$$\frac{3506}{1000} = x = 3.51$$

Simplify the equation.

Therefore, to prepare 200 mL of 0.3 *M* sodium chloride solution, 3.51 grams of NaCl are dissolved in distilled water to a final volume of 200 mL.

Problem 2.9 How are 50 mL of 20 millimolar (m*M*) sodium hydroxide (NaOH) prepared?

Solution 2.9 First, because it is somewhat more convenient to deal with terms expressed as molarity (*M*), convert the 20 m*M* value to an *M* value:

$$20 \text{ m}M \times \frac{1\ M}{1000 \text{ m}M} = 0.02\ M$$

Next, set up a ratio to calculate the amount of NaOH (40.0 gram molecular weight) needed to prepare 1 L of 0.02 *M* NaOH. Use the expression "40.0 grams is to 1 *M* as *x* grams is to 0.02 *M*." Solve for *x*.

$$\frac{40.0 \text{ g}}{1\ M} = \frac{x \text{ g}}{0.02\ M}$$

$$\frac{(40.0)(0.02)}{1} = x$$

Multiply both sides of the equation by 0.02.

$$0.8 = x$$

Therefore, if 1 L of 0.02 *M* NaOH is to be prepared, add 0.8 grams of NaOH to water to a final volume of 1000 mL.

Now, set up a ratio to determine how much is required to prepare 50 mL and solve for x. (The relationship of ratios should read as follows: 0.8 grams is to 1000 mL as x grams is to 50 mL.)

$$\frac{0.8 \text{ g}}{1000 \text{ mL}} = \frac{x \text{ g}}{50 \text{ mL}}$$

$$\frac{(0.8)(50)}{1000} = \frac{(x)(50)}{50}$$

Multiply both sides of the equation by 50. This cancels the 50 in the denominator on the right side of the equation.

$$\frac{40.0}{1000} = x = 0.04$$

Simplify the equation.

Therefore, to prepare 50 mL of 20 m*M* NaOH, 0.04 grams of NaOH is dissolved in a final volume of 50 mL of distilled water.

Problem 2.10 How many moles of NaCl are present in 50 mL of a 0.15 *M* solution?

Solution 2.10 A 0.15 *M* solution of NaCl contains 0.15 moles of NaCl per liter. Since we want to know how many moles are in 50 mL, we need to use a conversion factor to convert liters to milliliters. The equation can be written as follows:

$$x \text{ mol} = 50 \text{ mL} \times \frac{1 \text{ L}}{1000 \text{ mL}} \times \frac{0.15 \text{ mol}}{\text{L}}$$

Notice that terms on the right side of the equation cancel except for moles. Multiplying numerator and denominator values gives

$$x \text{ mol} = \frac{(50) \times (1) \times (0.15) \text{ mol}}{1000} = \frac{7.5 \text{ mol}}{1000} = 0.0075 \text{ mol}$$

Therefore, 50 mL of 0.15 *M* NaCl contains 0.0075 moles of NaCl.

Diluting Molar Solutions

Diluting stock solutions prepared in molar concentration into volumes of lesser molarity is performed as previously described (see page 18).

Problem 2.11 From a 1 *M* Tris solution, how is a 400 mL of 0.2 *M* Tris prepared?

Solution 2.11 The following equation is used to solve for *x*, the amount of 1 *M* Tris added to 400 mL to yield a 0.2 *M* solution.

$$1\ M \times \frac{x\ \text{mL}}{400\ \text{mL}} = 0.2\ M$$

$$\frac{1x\ M}{400} = 0.2\ M$$

The mL units cancel since they appear in both the numerator and the denominator on the left side of the equation. Multiply numerator values.

$$x = (0.2)(400)$$

Multiply both sides of the equation by 400.

$$x = 80$$

Therefore, to 320 mL (400 mL – 80 mL = 320 mL) of distilled water, add 80 mL of 1 *M* Tris, pH 8.0, to bring the solution to 0.2 *M* and a final volume of 400 mL.

Problem 2.12 How are 4 mL of 50 m*M* NaCl solution prepared from a 2 *M* NaCl stock?

Solution 2.12 In this example, a conversion factor must by included in the equation so that molarity (*M*) can be converted to millimolarity.

$$2\ M \times \frac{1000\ \text{m}M}{M} \times \frac{x\ \text{mL}}{4\ \text{mL}} = 50\ \text{m}M$$

$$\frac{2000x\ \text{m}M}{4} = 50\ \text{m}M$$

Multiply numerators. (On the left side of the equation, the *M* and mL units both cancel since these terms appear in both
the numerator and the denominator.)

$$2000x\ \text{mM} = 200\ \text{m}M$$

Multiply both sides of the equation by 4.

$$x = \frac{200 \text{ m}M}{2000 \text{ m}M} = 0.1$$

Divide each side of the equation by 2000 m*M*.

Therefore, to 3.9 mL of distilled water, add 0.1 mL of 2 *M* NaCl stock solution to produce 4 mL final volume of 50 m*M* NaCl.

Converting Molarity to Percent

Since molarity is a concentration of grams per 1000 mL, it is a simple matter to convert it to a percent value, an expression of a gram amount in 100 mL. The method is demonstrated in the following problem.

Problem 2.13 Express 2.5 *M* NaCl as a percent solution.

Solution 2.13 The gram molecular weight of NaCl is 58.44. The first step in solving this problem is to determine how many grams of NaCl are in a 2.5 *M* NaCl solution. This can be accomplished by using an equation of ratios: "58.44 grams is to 1 *M* as *x* grams is to 2.5 *M*." This relationship is expressed mathematically as follows:

$$\frac{58.44 \text{ g}}{1 \ M} = \frac{x \text{ g}}{2.5 \ M}$$

Solve for *x*.

$$\frac{(58.44)(2.5)}{1} = x$$

Multiply both sides of the equation by 2.5.

$$146.1 = x$$

Simplify the equation.

Therefore, to prepare a 2.5 *M* solution of NaCl, 146.1 grams of NaCl are dissolved in a total volume of 1 liter.

Percent is an expression of concentration in parts per 100. To determine the relationship between the number of grams of NaCl present in a 2.5 *M* NaCl solution and the equivalent percent concentration, ratios can be set up that state, "146.1 g is to 1000 mL as *x* g is to 100 mL."

$$\frac{146.1 \text{ g}}{1000 \text{ mL}} = \frac{x \text{ g}}{100 \text{ mL}}$$

Solve for *x*.

$$\frac{(146.1)(100)}{1000} = x$$

Multiply both sides of the equation by 100 (the denominator on the right side of the equation).

$$\frac{14610}{1000} = x = 14.6$$

Simplify the equation.

Therefore, a 2.5 *M* NaCl solution contains 14.6 grams of NaCl in 100 mL, which is equivalent to a 14.6% NaCl solution.

Converting Percent to Molarity

Converting a solution expressed as percent to one expressed as a molar concentration is a matter of changing an amount per 100 mL to an equivalent amount per liter (1000 mL) as demonstrated in the following problem.

Problem 2.14 What is the molar concentration of a 10% NaCl solution?

Solution 2.14 A 10% solution of NaCl, by definition, contains 10 grams of NaCl in 100 mL of solution. The first step to solving this problem is to calculate the amount of NaCl in 1000 mL of a 10% solution. This is accomplished by setting up a relationship of ratios as follows:

$$\frac{10 \text{ g}}{100 \text{ mL}} = \frac{x \text{ g}}{1000 \text{ mL}}$$

Solve for x.

$$\frac{(10)(1000)}{100} = x$$

Multiply both sides of the equation by 1000 (the denominator on the right side of the equation).

$$\frac{10{,}000}{100} = x = 100$$

Simplify the equation.

Therefore, a 1000-mL solution of 10% NaCl contains 100 grams of NaCl.

Using the gram molecular weight of NaCl (58.44), an equation of ratios can be written to determine molarity. In the following equation, we determine the molarity (*M*) equivalent to 100 grams:

$$\frac{x\ M}{100\ \text{g}} = \frac{1\ M}{58.44\ \text{g}}$$

Solve for x.

$$x = \frac{100}{58.44} = 1.71$$

Multiply both sides of the equation by 100 and divide by 58.44.

Therefore, a 10% NaCl solution is equivalent to 1.71 M NaCl.

Normality

A 1 normal (1 N) solution is equivalent to the gram molecular weight of a compound divided by the number of hydrogen ions present in solution (i.e., dissolved in one liter of water). For example, the gram molecular weight of hydrochloric acid (HCl) is 36.46. Since, in a solution of HCl, one H^+ ion can combine with Cl^- to form HCl, a 1 N HCl solution contains 36.46/1 = 36.46 grams HCl in 1 liter. A 1 N HCl solution, therefore, is equivalent to a 1 M HCl solution. As another example, the gram molecular weight of sulfuric acid (H_2SO_4) is 98.0. Since, in a H_2SO_4 solution, two H^+ ions can combine with SO_4^{2-} to form H_2SO_4, a 1 N H_2SO_4 solution contains 98.0/2 = 49.0 grams of H_2SO_4 in 1 liter. Since half the gram molecular weight of H_2SO_4 is used to prepare a 1 N H_2SO_4 solution, a 1 N H_2SO_4 solution is equivalent to a 0.5 M H_2SO_4 solution.

Normality and molarity are related by the equation

$$N = nM$$

where n is equal to the number of replaceable H^+ (or Na^+) or OH^- ions per molecule.

Problem 2.15 What is the molarity of a 1 N sodium carbonate (Na_2CO_3) solution?

Solution 2.15 Sodium carbonate has two replaceable Na^+ ions. The relationship between normality and molarity is

$$N = nM$$

Solving for M gives the following result.

$$\frac{N}{n} = M$$

Divide both sides of the equation by n (the number of replaceable Na^+ ions).

$$\frac{1}{2} = M$$

In this problem, the number of replaceable + ions, n, is 2. The normality, N, is 1.

$$0.5 = M$$

Therefore, a 1 N sodium carbonate solution is equivalent to a 0.5 M sodium carbonate solution.

pH

The first chemical formula most of us learn, usually during childhood, is that for water, H_2O. A water molecule is composed of two atoms of hydrogen and one atom of oxygen. The atoms of water, however, are only transiently associated in this form. At any particular moment, a certain number of water molecules will be dissociated into hydrogen (H^+) and hydroxyl (OH^-) ions These ions will reassociate within a very short time to form the H_2O water molecule again. The dissociation and reassociation of the atoms of water can be depicted by the following relationship:

$$H_2O \rightleftarrows H^+ + OH^-$$

In actuality, the hydrogen ion is donated to another molecule of water to form a hydronium ion (H_3O^+). A more accurate representation of the dissociation of water, therefore, is as follows:

$$H_2O + H_2O \rightleftarrows H_3O^+ + OH^-$$

For most calculations in chemistry, however, it is simpler and more convenient to think of the H^+ hydrogen ion as being a dissociation product of H_2O rather than the H_3O^+ hydronium ion.

A measure of the hydrogen ion (H^+) concentration in a solution is given by a **pH** value. A solution's pH is defined as the negative logarithm to the base 10 of its hydrogen ion concentration:

$$pH = -\log[H^+]$$

In this nomenclature, the brackets signify concentration. A **logarithm (log)** is an exponent, a number written above, smaller, and to the right of another number, called the **base**, to which the base should be raised. For example, for 10^2, the 2 is the exponent and the 10 is the base. In 10^2, the base, 10, should be raised to the second power, which means that 10 should be multiplied by itself a total of two times. This will give a value of 100, as shown here:

$$10^2 = 10 \times 10 = 100$$

The log of 100 is 2 because that is the exponent of 10 that yields 100. The log of 1000 is 3 since $10^3 = 1000$:

$$10^3 = 10 \times 10 \times 10 = 1000$$

The log of a number can be found on most calculators by entering the number and then pressing the **log** key.

In pure water, the H^+ concentration ($[H^+]$) is equal to 10^{-7} *M*. In other words, in 1 L of water, 0.0000001 moles of hydrogen ion will be present. The pH of water, therefore, is calculated as follows:

$$\text{pH} = -\log(10^{-7}) = -(-7) = 7$$

Pure water, therefore, has a pH of 7.0.

pH values range from 0 to 14. Solutions having pH values less than 7 are acidic. Solutions having pH values greater than 7 are alkaline, or basic. Water, with a pH of 7, is considered a neutral solution; it is neither acidic nor basic.

When an acid, such as hydrochloric acid (HCl), is added to pure water, the hydrogen ion concentration increases above 10^{-7} *M*. When a base, such as sodium hydroxide (NaOH), is added to pure water, the OH^- ion is dissociated from the base. This hydroxyl ion can associate with the H^+ ions already in the water to form H_2O molecules, reducing the solution's hydrogen ion concentration and increasing the solution's pH.

Problem 2.16 The concentration of hydrogen ion in a solution is 10^{-5} *M*. What is the solution's pH?

Solution 2.16 pH is the negative logarithm of 10^{-5}.

$$\text{pH} = -\log[H^+] = -\log(10^{-5}) = -(-5) = 5$$

Therefore, the pH of the solution is 5. It is acidic.

Problem 2.17 The concentration of hydrogen ion in a solution is 2.5×10^{-4} M. What is the solution's pH?

Solution 2.17 The **product rule for logarithms** states that for any positive numbers M, N, and a (where a is not equal to 1), the logarithm of a product is the sum of the logarithms of the factors:

$$\log_a MN = \log_a M + \log_a N$$

Since we are working in base 10, a is 10.

The product rule of logarithms will be used to solve this problem.

$$\begin{aligned} \text{pH} &= -\ \log\left(2.5 \times 10^{-8}\right) \\ &= -\ \left(\log 2.5 + \log 10^{-5}\right) \\ &= -\left[0.40 + (-5)\right] \\ &= -(0.40 - 5) = -(-4.6) = 4.6 \end{aligned}$$

Therefore, the solution has a pH of 4.6.

Note: *In Problem 2.16, the hydrogen ion concentration was stated to be 10^{-5}. This value can also be written as 1×10^{-5}. The log of 1 is 0. If the product rule for logarithms is used to calculate the pH for this problem, it would be equal to −[0 +(−5)], which is equal to 5.*

Problem 2.18 The pH of a solution is 3.75. What is the concentration of hydrogen ion in the solution?

Solution 2.18 To calculate the hydrogen ion concentration in this problem will require that we determine the **antilog** of the pH. An antilog is found by doing the reverse process of that used to find a logarithm. The log of 100 is 2. The antilog of 2 is 100. The log of 1000 is 3. The antilog of 3 is 1000. For those calculators that do not have an antilog key, this can usually be obtained by entering the value, pressing the $\mathbf{10^x}$ key, and then pressing the = sign. (Depending on the type of calculator you are using, you may need to press the **SHIFT** key to gain access to the $\mathbf{10^x}$ function.)

$\text{pH} = -\log\,[H^+]$ Equation for calculating pH.

$-\log[H^+] = 3.75$ The pH is equal to 3.75.

$\log[H^+] = -3.75$ Multiply each side of the equation by −1.

$[H^+] = 1.8 \times 10^{-4}$

Take the antilog of each side of the equation.

***Note:** Taking the antilog of the log of a number, since they are opposite and canceling operations, is equivalent to doing nothing to that number. For example, the antilog of the log of 100 is 100.*

Therefore, the hydrogen ion concentration is 1.8×10^{-4} *M*.

Since water, H_2O, dissociates into both H^+ and OH^- ions, the H^+ concentration must equal the OH^- concentration. Just as water has a pH, so does it have a **pOH**, which is defined as the negative logarithm of the OH^- (hydroxyl ion) concentration.

$$pOH = -\log[OH]$$

and

$$pOH = 14 - pH$$

Problem 2.19 A solution has a pH of 4.5. What is the solution's pOH?

Solution 2.19 The pOH is obtained by subtracting the pH from 14.

$$pOH = 14 - pH$$

$$pOH = 14 - 4.5 = 9.5$$

Therefore, the pOH of the solution is 9.5.

Problem 2.20 What is the pH of a 0.02 *M* solution of sodium hydroxide (NaOH).

Solution 2.20 Sodium hydroxide is a strong base and, as such, is essentially ionized completely to Na^+ and OH^- in dilute solution. The OH^- concentration, therefore, is 0.02 *M*, the same as the concentration of NaOH. For a strong base, the H^+ ion contribution from water is negligible and so will be ignored. The first step to solving this problem is to determine the pOH. The pOH value will then be subtracted from 14 to obtain the pH.

$$\text{pOH} = -\log(0.02)$$
$$= -(-1.7) = 1.7$$
$$\text{pH} = 14 - 1.7 = 12.3$$

Therefore, the pH of the 0.02 *M* NaOH solution is 12.3.

pK_a and the Henderson–Hasselbalch Equation

In the Bronsted concept of acids and bases, an **acid** is defined as a substance that donates a proton (a hydrogen ion). A **base** is a substance that accepts a proton. When a Bronsted acid loses a hydrogen ion, it becomes a Bronsted base. The original acid is called a **conjugate acid**. The base created from the acid by loss of a hydrogen ion is called a **conjugate base**.

Dissociation of an acid in water follows the general formula

$$HA + H_2O \rightleftharpoons H_3O^+ + A^-$$

Where HA is a conjugate acid, H_2O is a conjugate base, H_3O^+ is a conjugate acid, and A^- is a conjugate base.

The acid's ionization can be written as a simple dissociation, as follows:

$$HA \rightleftharpoons H^+ + A^-$$

The dissociation of the HA acid will occur at a certain rate characteristic of the particular acid. Notice, however, that the arrows go in both directions. The acid dissociates into its component ions, but the ions come back together again to form the original acid. When the rate of dissociation into ions is equal to the rate of ion reassociation, the system is said to be in **equilibrium**. A strong acid will reach equilibrium at the point where it is completely dissociated. A weak acid will have a lower percentage of molecules in a dissociated state and will reach equilibrium at a point less than 100% ionization. The concentration of acid at which equilibrium occurs is called the **acid dissociation constant**, designated by the symbol K_a. It is represented by the following equation:

$$K_a = \frac{[H^+][A^-]}{[HA]}$$

A measure of K_a for a weak acid is given by its pK_a, which is equivalent to the negative logarithm of K_a:

$$\mathrm{p}K_a = -\log K_a$$

pH is related to pK_a by the **Henderson–Hasselbalch equation**:

$$\mathrm{pH} = \mathrm{p}K_a + \log\frac{[\text{conjugate base}]}{[\text{acid}]}$$

$$= \mathrm{p}K_a + \log\frac{[\mathrm{A}^-]}{[\mathrm{HA}]}$$

The Henderson–Hasselbalch equation can be used to calculate the amount of acid and conjugate base to be combined for the preparation of a buffer solution having a particular pH, as demonstrated in the following problem.

Problem 2.21 You wish to prepare 2 liters of 1 *M* sodium phosphate buffer, pH 8.0. You have stocks of 1 *M* monobasic sodium phosphate (NaH_2PO_4) and 1 *M* dibasic sodium phosphate (Na_2HPO_4). How much of each stock solution should be combined to make the desired buffer?

Solution 2.21 Monobasic sodium phosphate (NaH_2PO_4) in water exists as Na^+ and $H_2PO_4^-$ ions. $H_2PO_4^-$ (phosphoric acid, the conjugate acid) dissociates further to HPO_4^{2-} (the conjugate base) + H^+ and has a pK_a of 6.82 at 25°C. [pK_a values can be found in the Sigma chemical catalogue (Sigma, St. Louis, MO) or in *The CRC Handbook of Chemistry and Physics*.] The pH and pK_a values will be inserted into the Henderson–Hasselbalch equation to derive a ratio of the conjugate base and acid to combine to give a pH of 8.0. Note that the stock solutions are both at a concentration of 1 *M*. No matter in what ratio the two solutions are combined, there will always be 1 mole of phosphate molecules per liter.

$$\mathrm{pH} = \mathrm{p}K_a + \log\frac{[\mathrm{A}^-]}{[\mathrm{HA}]}$$

Insert pH and pK_a values into Henderson–Hasselbalch equation.

$$8.0 = 6.82 + \log\frac{[\mathrm{HPO_4}^{2-}]}{[\mathrm{H_2PO_4}^-]}$$

$$1.18 = \log\frac{[\mathrm{HPO_4}^{2-}]}{[\mathrm{H_2PO_4}^-]}$$

Subtract 6.82 from both sides of the equation.

$$\text{antilog } 1.18 = \frac{[\mathrm{HPO_4}^{2-}]}{[\mathrm{H_2PO_4}^-]} = 15.14$$

Take the antilog of each side of the equation.

Therefore, the ratio of HPO_4^{2-} to $H_2PO_4^-$ is equal to 15.14. To make 1 *M* sodium phosphate buffer, 15.14 parts Na_2HPO_4 should be combined with 1 part NaH_2PO_4. 15.14 parts Na_2HPO_4 plus 1 part NaH_2PO_4 is equal to a total of 16.14 parts. The amounts of each stock to combine to make 2 liters of the desired buffer is then calculated as follows:

For Na_2HPO_4, the amount is equal to

$$\frac{15.14}{16.14} \times 2 \text{ L} = 1.876 \text{ L}$$

For NaH_2PO_4, the amount is equal to

$$\frac{1}{16.14} \times 2 \text{ L} = 0.124 \text{ L}$$

When these two volumes are combined, you will have 1 *M* sodium phosphate buffer having a pH of 8.0.

Cell Growth

3

The Bacterial Growth Curve

Where bacteria or other unicellular organisms are concerned, **cell growth** refers to cell division and the increase in cell quantity rather than to the size of an individual cell. It's about numbers. The rate at which bacterial cells divide in culture is influenced by several factors, including the level of available nutrients, the temperature at which the cells are incubated, and the degree of aeration. Depending on their genotypes, different strains of bacteria may also have different growth rates in any particular defined medium.

An understanding of the characteristics of cell growth is important for several applications in molecular biology and biotechnology:

- For the most efficient and reproducible transformations of *E. coli* by recombinant plasmid, it is important that the cells be harvested and prepared at a particular point in their growth, at a particular cell concentration.

- *E. coli* is most receptive to bacteriophage infection during a certain period of growth.

- The highest yield of a cellular or recombinant protein can be achieved during a particular window of cell growth.

A common procedure for determining the rate of cell growth is to start with a large volume (50 to 100 mL) of a defined medium containing a small inoculum (1 mL) of cells from a liquid culture that had been grown over the previous night. The point at which the larger culture is inoculated is considered to be time zero. The culture is then incubated at the proper temperature (37°C with aeration for most laboratory strains). At various times, a sample of the culture is withdrawn and its optical density (OD) is determined. That sample is then diluted to a concentration designed to give easily countable, well-isolated colonies when spread onto petri plates containing a solid agar-based medium. The plates are incubated overnight and the number of colonies is counted. If colonies on the plate are well separated, it can be assumed that each colony arises from a single viable cell. The OD at the various sampling times can then be correlated with cell number and a growth rate can be derived. For subsequent experiments, if all conditions are carefully reproduced, the number of cells present at any particular time during incubation can be extrapolated by measuring the culture's OD.

As an example of the type of math involved in this procedure, imagine that a growth experiment has been set up as just described in which a 100-mL volume of tryptone broth is inoculated with 0.5 mL of overnight culture of *E. coli*. At the time of inoculation, a 0.75-mL sample is withdrawn and its optical density at 550 nm (OD_{550}) is determined. (Beyond a certain OD_{550} reading, an increase in cell density will no longer result in a linear increase in absorbance. For many spectrophotometers, this value is an OD_{550} of 1.5. Extremely dense cultures, therefore, should be diluted to keep the absorbance at 550 nm below 1.5.)

Following the reading of a sample's absorbance, a **serial dilution** is then performed by transferring 0.1 mL from the 0.75-mL sample into 9.9 mL of tryptone broth. That diluted sample is vortexed, and 0.2 mL is withdrawn and diluted into a second tube, containing 9.8 mL of tryptone broth. The second diluted sample is vortexed to ensure proper mixing and 0.1 mL is withdrawn and spread onto an agar plate. Following overnight incubation of that plate, 420 colonies are counted.

Problem 3.1 In the example just given, what is the dilution of cells in the second tube?

Solution 3.1 The dilutions can be represented as fractions. A 0.1-mL aliquot diluted into 9.9 mL can be written as the fraction 0.1/10, where the numerator is the volume of sample being transferred (0.1 mL) and the denominator is equal to the final volume of the dilution (0.1 mL + 9.9 mL = 10.0 mL). Multiple dilutions are multiplied together to give the dilution in the final sample. This approach can be simplified if the fractions are converted to scientific notation (see Chapter 1 for a discussion of how to convert fractions to scientific notation):

$$\frac{0.1\ \text{mL}}{10\ \text{mL}} \times \frac{0.2\ \text{mL}}{10\ \text{mL}} = \left(1 \times 10^{-2}\right)\left(2 \times 10^{-2}\right) = 2 \times 10^{-4}$$

Therefore, in the sample in the second tube has been diluted to 2×10^{-4}.

Problem 3.2 What is the dilution of cells spread onto the agar plate?

Solution 3.2 One-tenth mL of the dilution is spread on the plate. This gives

$$\frac{0.1\ \text{mL}}{10\ \text{mL}} \times \frac{0.2\ \text{mL}}{10\ \text{mL}} \times 0.1\ \text{mL} = \left(1 \times 10^{-2}\right)\left(2 \times 10^{-2}\right)\left(1 \times 10^{-1}\ \text{mL}\right) = 2 \times 10^{-5}\ \text{mL}$$

Therefore, a dilution of 2×10^{-5} mL is spread on the agar plate. Note that this amount is equivalent to withdrawing 0.00002 mL (20 nanoliters) directly from the 100-mL culture for plating. Performing dilutions, therefore, allows the experimenter to withdraw, in manageable volumes, what is essentially a very small amount of a solution.

Problem 3.3 What is the concentration of viable cells in the 100-mL culture immediately following inoculation?

Solution 3.3 Each colony on the spread plate can be assumed to have arisen from a viable cell. To determine the number of viable cells in the 100-mL culture, the number of colonies counted on the spread plate following overnight incubation is divided by the dilution:

$$\frac{420 \text{ cells}}{2\times10^{-5} \text{ mL}} = \frac{4.2\times10^{2} \text{ cells}}{2\times10^{-5} \text{ mL}} = 2.1\times10^{7} \text{ cells/mL}$$

Therefore, there are 2.1×10^{7} cells/mL in the100-mL culture after inoculation.

Problem 3.4 How many total viable cells are present in the100-mL culture?

Solution 3.4 Multiply the cell concentration by the culture volume:

$$\frac{2.1\times10^{7} \text{ cells}}{\text{mL}} \times 100 \text{ mL} = 2.1\times10^{9} \text{ total cells}$$

Therefore, there are 2.1×10^{9} viable cells in the 100-mL culture.

Problem 3.5 What is the concentration of cells in the overnight culture from which the 0.5-mL sample was withdrawn and used for inoculating the 100 mL of growth media? (This problem can be solved by two methods.

Solution 3.5(a) In Problem 3.4, it was determined that the 100-mL culture, at time zero, contained 2.1×10^{9} cells. Those cells came from 0.5 mL of the overnight culture, or

$$\frac{2.1\times10^{9} \text{ cells}}{0.5 \text{ mL}} = \frac{2.1\times10^{9} \text{ cells}}{5\times10^{-1} \text{ mL}} = 4.2\times10^{9} \text{ cells/mL}$$

Solution 3.5(b) An equation can be written that describes the dilution of the overnight culture [0.5 mL of cells of unknown concentration (x cells/mL) transferred into 100 mL of fresh tryptone broth to give a concentration of 2.1×10^{7} cells/mL]:

$$\frac{x \text{ cells}}{\text{mL}} \times \frac{0.5 \text{ mL}}{100 \text{ mL}} = \frac{2.1\times10^{7} \text{ cells}}{\text{mL}}$$

Solve for x.

In the 0.5 mL/100 mL fraction, the mL terms cancel since they are present in both the numerator and the denominator. Multiplying numerators on the left side of the equation and multiplying both sides of the equation by 100 (written in scientific notation as 1×10^{2}) gives

$$\frac{0.5x \text{ cells}}{\text{mL}} = \frac{(2.1\times10^{7} \text{ cells})(1\times10^{2})}{\text{mL}} = \frac{2.1\times10^{9} \text{ cells}}{\text{mL}}$$

Dividing each side of the equation by 0.5 (written in scientific notation as 5×10^{-1}) yields the following equation:

$$\frac{x \text{ cells}}{\text{mL}} = \frac{2.1\times10^{9} \text{ cells}}{5\times10^{-1} \text{ mL}} = \frac{0.42\times10^{9-(-1)} \text{ cells}}{\text{mL}} = \frac{4.2\times10^{9} \text{ cells}}{\text{mL}}$$

Therefore, the overnight culture has a concentration of 4.2×10^{9} cells/mL.

Sample Data

For the example just described, at 1-hour intervals, 0.75-mL samples are withdrawn, their optical densities are determined, and each sample is diluted and plated to determine viable cell count. The data set in Table 3.1 is obtained.

Hours After Inoculation	OD_{550}	Cells/mL
0	0.008	2.1×10^{7}
1	0.020	4.5×10^{7}
2	0.052	1.0×10^{8}
3	0.135	2.6×10^{8}
4	0.200	4.0×10^{8}
5	0.282	5.8×10^{8}
6	0.447	9.6×10^{8}
7	0.661	1.5×10^{9}
8	1.122	2.0×10^{9}

Table 3.1 Sample data for Problems 3.6 through 3.15.

Manipulating Cell Concentration

The following problems demonstrate how cell cultures can be manipulated by dilutions to provide desired cell concentrations.

Problem 3.6 At $t = 3$ hours, the cell density was determined to be 2.6×10^8 cells/mL. If 3.5 mL of culture are withdrawn at that time, the aliquot is centrifuged to pellet the cells, and the pellet is then resuspended in 7 mL of tryptone broth, what is the new cell concentration?

Solution 3.6 It is first necessary to determine the number of cells in the 3.5-mL aliquot. This is done by multiplying the cell concentration (2.6×10^8 cells/mL) by the amount withdrawn (3.5 mL). The following equation is used to solve for x, the number of cells in 3.5 mL:

$$\frac{2.6 \times 10^8 \text{ cells}}{\text{mL}} \times 3.5 \text{ mL} = x \text{ cells}$$

Solving for x yields

$$\frac{(2.6 \times 10^8 \text{ cells}) \times (3.5 \text{ m}L)}{\text{mL}} = x \text{ cells}$$

$$9.1 \times 10^8 \text{ cells} = x \text{ cells}$$

Therefore, there are 9.1×10^8 cells in the 3.5-mL aliquot. When these 9.1×10^8 cells are resuspended in 7 mL, the new concentration can be calculated as follows:

$$\frac{x \text{ cells}}{\text{mL}} = \frac{9.1 \times 10^8 \text{ cells}}{7 \text{ mL}} = \frac{1.3 \times 10^8 \text{ cells}}{\text{mL}}$$

Therefore, the cells resuspended in 7 mL of tryptone broth have a concentration of 1.3×10^8 cells/mL.

Problem 3.7 In Problem 3.6, 3.5 mL of culture at 2.6×10^8 cells/mL are pelleted by centrifugation. Into what volume should the pellet be resuspended to obtain a cell concentration of 4.0×10^9 cells/mL?

Solution 3.7 In Problem 3.6, it was determined that there are a total of 9.1×10^8 cells in a 3.5-mL sample of a culture having a concentration of 2.6×10^8 cells/mL. The following equation can be used to determine the volume required to resuspend the centrifuged pellet to obtain a concentration of 4.0×10^9 cells/mL:

$$\frac{9.1\times10^8 \text{ cells}}{x \text{ mL}} = \frac{4.0\times10^9 \text{ cells}}{\text{mL}}$$

Solve for x.

$$9.1\times10^8 \text{ cells} = \frac{(4.0\times10^9 \text{ cells})\times(x \text{ mL})}{\text{mL}}$$

Multiply both sides of the equation by x.

$$\frac{9.1\times10^8 \text{ cells}}{4.0\times10^9 \text{ cells}} = x = 0.23$$

Divide each side of the equation by 4.0×10^9 cells.

Therefore, to obtain a concentration of 4.0×10^9 cells/mL, the pelleted cells should be resuspended in 0.23 mL of tryptone broth.

Problem 3.8 At 6 hours following inoculation, the cell concentration is 9.6×10^8 cells/mL. If a sample is withdrawn at this time, how should it be diluted to give 300 colonies when spread on an agar plate?

Solution 3.8 Since each colony on a plate is assumed to arise from an individual cell, the first step to solving this problem is to determine how many milliliters of culture contains 300 cells. The following equation will provide this answer. The equation is written so that the mL terms cancel.

$$\frac{9.6\times10^8 \text{ cells}}{\text{mL}} \times x \text{ mL} = 300 \text{ cells}$$

Solving for x gives

$$x = \frac{3\times10^2 \text{ cells}}{9.6\times10^8 \text{ cells}} = 3.1\times10^{-7}$$

Therefore, 300 cells are contained in 3.1×10^{-7} mL of culture at $t = 6$ hours. Assume that 0.1 mL of the dilution series to be made will be plated. To perform an equivalent dilution to 2.1×10^{-7} mL, the following serial dilution can be performed.

$$(1\times10^{-2})(1\times10^{-2})(1\times10^{-1})(3.1\times10^{-1})(1\times10^{-1}) = 3.1\times10^{-7}$$

This is equivalent to performing the following dilution series and spreading 0.1 mL on an agar plate:

$$\frac{0.1\ \text{mL}}{10\ \text{mL}} \times \frac{0.1\ \text{mL}}{10\ \text{mL}} \times \frac{1.0\ \text{mL}}{10\ \text{mL}} \times \frac{3.1\ \text{mL}}{10\ \text{mL}} \times 0.1\ \text{mL}$$

Note: *There is no one correct way to determine a dilution series that will achieve the dilution you want. The volumes you use for the dilutions are dictated by several parameters, including the amount of dilution buffer available, the accuracy of the pipets, and the size of the available dilution tubes.*

Plotting OD_{550} vs. Time on a Linear Graph

A linear plot of the Table 3.1 sample data, in which OD_{550} is placed on the vertical (y) axis and time on the horizontal (x) axis, will produce the curve shown in Figure 3.1.

When cells from an overnight culture are inoculated into fresh medium, as described in this chapter, they experience a period during which they adjust to the new environment. Very little cell division occurs during this time. This period, called the **lag phase**, may last from minutes to hours, depending on the strain and the richness of the medium. The culture then enters a period during which rapid cell division occurs. During this period, each cell in the population gives rise to two daughter cells. These two daughter cells give rise to four cells, those four to eight, etc. Because of the continuous doubling in cell number, this period of growth is called the **exponential** or **logarithmic (log) phase**. As the supply of nutrients is consumed and as inhibitory waste products accumulate, the rate of cell division slows. At this point, the culture enters the **stationary phase**, during which time cell death and cell growth occur at equivalent rates. The overall viable cell number during this stage remains constant. (For the data plotted in the experiment outlined earlier, the graph is seen to be entering this period.) Further depletion of nutrients leads to increased cell death. As more cells die than divide, the culture enters the **death phase**. Most strains have entered the cell death phase by 24 hours.

The phases of cell growth are not easily distinguishable in the plot in Figure 3.1. They become more discernable in a logarithmic plot (see next section).

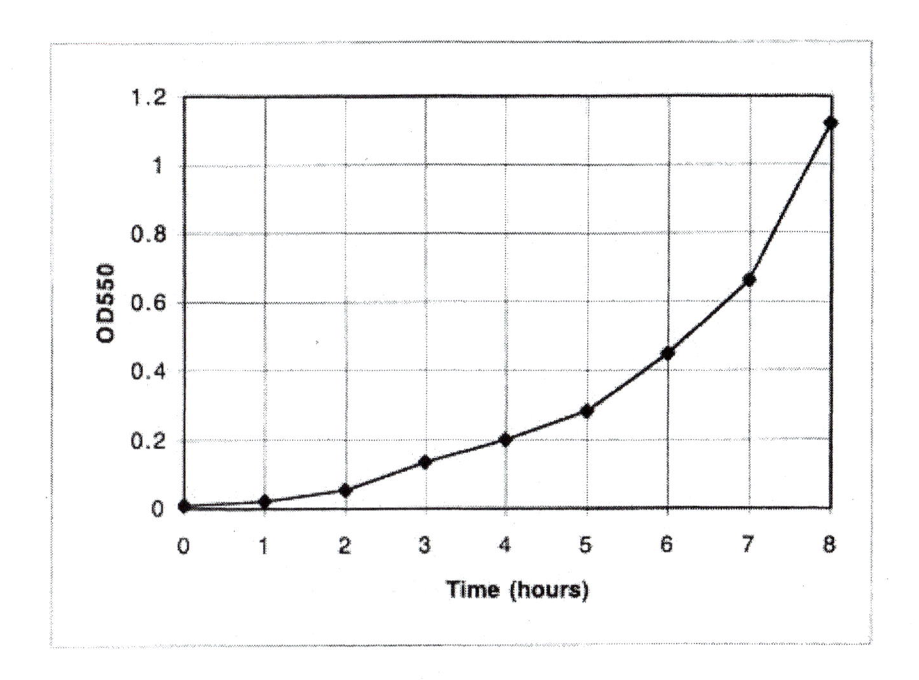

Figure 3.1. Plot of OD_{550} vs. time for sample data from Table 3.1.

Plotting the Logarithm of OD_{550} vs. Time on a Linear Graph

Logarithms

An **exponent** is a number written above and to the right of (and smaller than) another number, called the **base**, to indicate the power to which the base is to be raised. A **logarithm (log)** is an exponent. Most mathematics in molecular biology and in most basic sciences are performed using base 10, the decimal system for naming numbers (from 0 to 9) at positions around the decimal point. Since this is the "commonly used" system, logarithms of base-10 numbers are called **common logarithms**, and, unless specified otherwise, the base for a logarithm is assumed to be 10. The log, base 10, of a number *n* is the exponent to which 10 must be raised in order to get *n*. For example, $10^2 = 100$. Here, the exponent is 2, the base is 10. The log of 100, therefore, is equal to the exponent 2. The log of 1000 is 3, since $10^3 = 1000$. The log of 10 is 1, since $10^1 = 10$. The log of 20 is 1.3, since $10^{1.3} = 20$. Log values can be found in a log table or, on a calculator, by entering the *n* value and then pressing the ***log*** key. Plotting the logarithm of the OD_{550} readings of the bacterial culture versus time should result in a linear plot for those readings taken during the logarithmic phase of cell growth.

Sample OD_{550} Data Converted to Log Values

From the sample data (Table 3.1), the log values obtained for the OD_{550} readings are as given in Table 3.2.

Hours After Inoculation	OD_{550}	Log of OD_{550}
0	0.008	–2.10
1	0.020	–1.70
2	0.052	–1.28
3	0.135	–0.87
4	0.200	–0.70
5	0.282	–0.55
6	0.447	–0.35
7	0.661	–0.18
8	1.122	0.05

Table 3.2 OD_{550} readings from Table 3.1 converted to log values.

Plotting log OD_{550} vs. Time

When the log of the OD_{550} readings versus time are plotted, the curve in Figure 3.2 is generated.

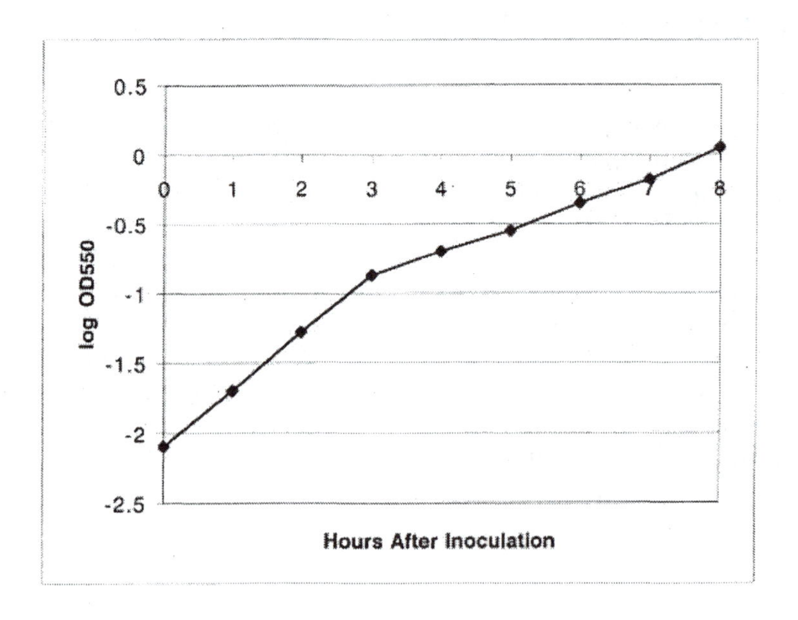

Figure 3.2 Plot of log OD_{550} vs. incubation time for sample data.

In Figure 3.2, the plot is linear from 0 to 3 hours following inoculation. This is the period of exponential growth. An inflection in the line occurs between 3 and 4 hours. This represents the end of exponential growth and the very beginning of stationary phase. This type of plot can give a better visual representation of the length of the exponential phase of growth and the point at which growth enters the stationary phase than that provided by the linear plot of OD_{550} versus time shown in Figure 3.1.

Plotting the Log of Cell Concentration vs. Time

Plotting the logarithm of cell concentration versus time should give a plot similar to that obtained when plotting the log of OD_{550} versus time.

Determining Log Values

Log values can be obtained on a calculator by entering a number and then pressing the ***log*** key. The log values of the cell concentrations at various times after inoculation are shown in Table 3.3.

Hours After Inoculation	Cells/mL	Log of Cell Concentration
0	2.1×10^7	7.32
1	4.5×10^7	7.65
2	1.0×10^8	8.00
3	2.6×10^8	8.41
4	4.0×10^8	8.60
5	5.8×10^8	8.76
6	9.6×10^8	8.98
7	1.5×10^9	9.18
8	2.0×10^9	9.30

Table 3.3 Log of cell concentration for Table 3.1 sample data.

When the log of cell concentration versus hours after inoculation is plotted, the graph shown in Figure 3.3 is generated.

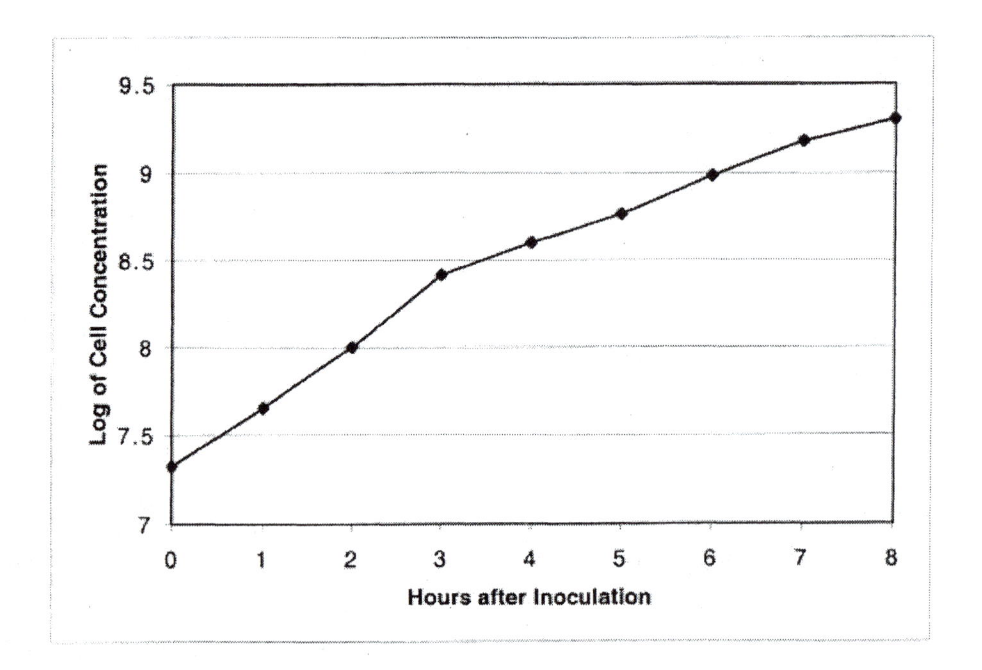

Figure 3.3 Data from Table 3.3 graphed on a linear plot.

Calculating Generation Time

Slope and the Growth Constant

The **generation time** of cells in culture is defined as the time required for the cell number to double. When the log of cell concentration versus time is plotted on linear graph paper, a line is generated rising from left to right. The vertical rise on the y axis divided by the horizontal length on the x axis over which the vertical rise occurs is the **slope** of the line. Slope, stated in a more popular way, is "the rise over the run." The slope of the line during exponential growth is designated as $\boldsymbol{K}$, the growth constant.

The increase in the number of cells (N) per unit time (t) is proportional to the number of cells present in the culture at the beginning of the time interval (N_0). This relationship is expressed by the formula

$$\log(N) = \log(N_0) + Kt$$

Subtracting $\log(N_0)$ from both sides of the equation and dividing both sides of the equation by t yields the slope of the line, K:

$$\frac{\log(N) - \log(N_0)}{t} = K$$

Problem 3.9 What is the slope (the K value) for the sample data in Table 3.1?

Solution 3.9 The preceding equation will be used. In the numerator, the log of an earlier cell concentration is subtracted from the log of a later cell concentration. An interval during the exponential phase of growth should be used. For example, since exponential growth is occurring during the time from 0 to 3 hours following inoculation, values obtained for the time interval from $t = 0$ to $t = 1$ or from $t = 1$ to $t = 2$ or from $t = 2$ to $t = 3$ can be used. A number closest to the actual K value can be obtained by using the values calculated from the $t = 0$ and $t = 3$ time points (a total of 180 minutes):

$$\frac{\log(N) - \log(N_0)}{t} = K$$

The cell concentration at $t = 3$ hours (180 minutes) is 2.6×10^8 cells/mL. The cell concentration at $t = 0$ is 2.1×10^7 cells/mL. These values are entered into the equation for calculating K.

$$\frac{\log(2.6 \times 10^8) - \log(2.1 \times 10^7)}{180 \text{ min}} = K$$

$$\frac{8.41 - 7.32}{180 \text{ min}} = K$$

Substitute the log values for N and N_0 into the equation.

$$\frac{1.09}{180 \text{ min}} = K = \frac{0.0061}{\text{min}}$$

Simplify the equation.

Therefore, the K value for this data set is 0.0061/min.

Generation Time

As shown earlier, the equation describing cell growth is given by the formula

$$\log(N) = \log(N_0) + Kt$$

The generation time of a particular bacterial strain under a defined set of conditions is the time required for the cell number to double. In this case, N would equal 2 and N_0 would equal 1 (or $N = 4$, $N_0 = 2$, or any other values representing a doubling in number). The formula then becomes

$$\log(2) = \log(1) + Kt$$

Converting to log values:

$$0.301 = 0 + Kt$$

Dividing each side of the equation by K gives the generation time t:

$$t = \frac{0.301}{K}$$

Problem 3.10 What is the generation time for the example culture?

Solution 3.10 From Problem 3.9, the growth constant, K, was calculated to be 0.0061/min. Placing this value into the expression for determining generation time gives

$$\text{generation time} = t = \frac{0.301}{K} = \frac{0.301}{0.0061/\text{min}} = 49.3 \text{ min}$$

Therefore, the generation time for the cells of this data set is 49.3 minutes.

Problem 3.11 What is the cell concentration 150 minutes after inoculation?

Solution 3.11 In Problem 3.9, a growth constant, K, of 0.0061/min was calculated. The initial cell concentration, 2.1×10^7 cells/mL, is found in Table 3.3. These values can be entered into the equation describing cell growth to determine the cell concentration at 150 minutes.

$$\log(N) = \log(N_0) + Kt$$

$$\log(N) = \log(2.1 \times 10^7) + \left(\frac{0.0061}{\text{min}}\right)(150 \text{ min})$$

The values for K, initial cell concentration, and t are placed into the equation.

$$\log(N) = 7.32 + 0.92 = 8.24$$

Simplify the equation.

To determine cell number, it must first be determined what number, when converted to a logarithm, is equal to 8.24. This number is the inverse of the log function and is called the **antilogarithm (antilog)** or **inverse logarithm**. The antilog of a number x (antilog x) is equal to 10^x.

Most calculators do not have a key marked "antilog." To find an antilog, you must use the **10^x** key. If this key does not exist, then you need to raise 10 to the *x* power using the **x^y** key. On some calculators, the **log** key serves as the **10^x** key after a **shift** or **inverse** key is pressed.

To find the antilog of 8.24 on a calculator, enter **8**, **.**, **2**, **4**, **shift**, and **10^x**. Or enter **1**, **0**, **x^y**, **8**, **.**, **2**, **4**, and **=**. Either path should yield an answer, when rounded off, of 1.7×10^8 cells/mL.

Therefore, at 150 minutes following inoculation, the culture will have a concentration of 1.7×10^8 cells/mL.

Plotting Cell Growth Data on a Semilog Graph

Semilog graph paper uses a log scale on the *y* (vertical) axis. The *y* axis is configured in "cycles." Each cycle is labeled 1 through 9 and corresponds to a power of 10. The scale between 1 and 2 is expanded compared with that from 2 to 3. The scale between 2 and 3 is expanded compared to that from 3 to 4. The divisions up the *y* axis become progressively smaller through 9, until 1 is reached and a new cycle begins. Semilog graph paper can be obtained commercially having as many as six cycles. Using semilog graph paper to plot growth data allows a quick visual determination of cell concentration at any point in time and an estimation of doubling time without having to perform log and antilog calculations. In the graphing methods outlined here, the sample data described in Table 3.1 will be used.

Plotting OD_{550} vs. Time on a Semilog Graph

Plotting OD_{550}, on the *y* axis, versus the hours following inoculation, on the *x* axis, is one way of representing the data. Since the OD_{550} values cover four places around the decimal (four decimal points), four-cycle semilog graph paper should be used (Figure 3.4).

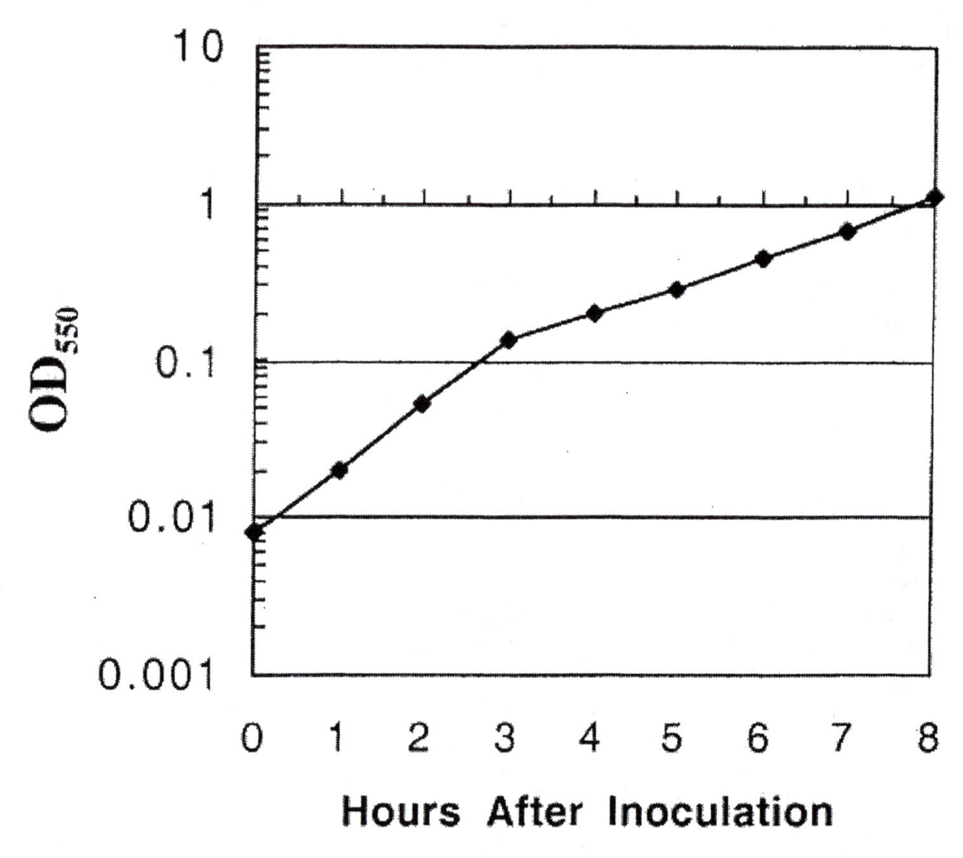

Figure 3.4 Plot of OD_{550} vs. time on a semilog graph.

Depending on how long following inoculation samples are taken, the curve may begin to level off (plateau) at later times as the OD_{550} value approaches 1.5. This can result from the culture's reaching its maximal absorbance level. Even though the cells may still be growing, the spectrophotometer is incapable of discerning a change.

Estimating Generation Time from a Semilog Plot of OD_{550} vs. Time
Generation time is also referred to as the **doubling time**. As shown in Problem 3.12, the generation time can be derived directly from a semilog plot by determining the time required for the OD_{550} value on the *y* axis to double.

Problem 3.12 Using a plot of OD_{550} vs. time, determine what the generation time is for those cells represented by the sample data.

Solution 3.12 Choose a point on the *y* axis (OD_{550}) in the region corresponding to the exponential growth phase. For example, an OD_{550} value of 0.03 can be chosen. Twice this value is 0.06. Draw horizontal lines from these two positions on the *y* axis to the plotted curve. From those intersection points on the curve, draw vertical lines to the *x* axis (Figure 3.5).

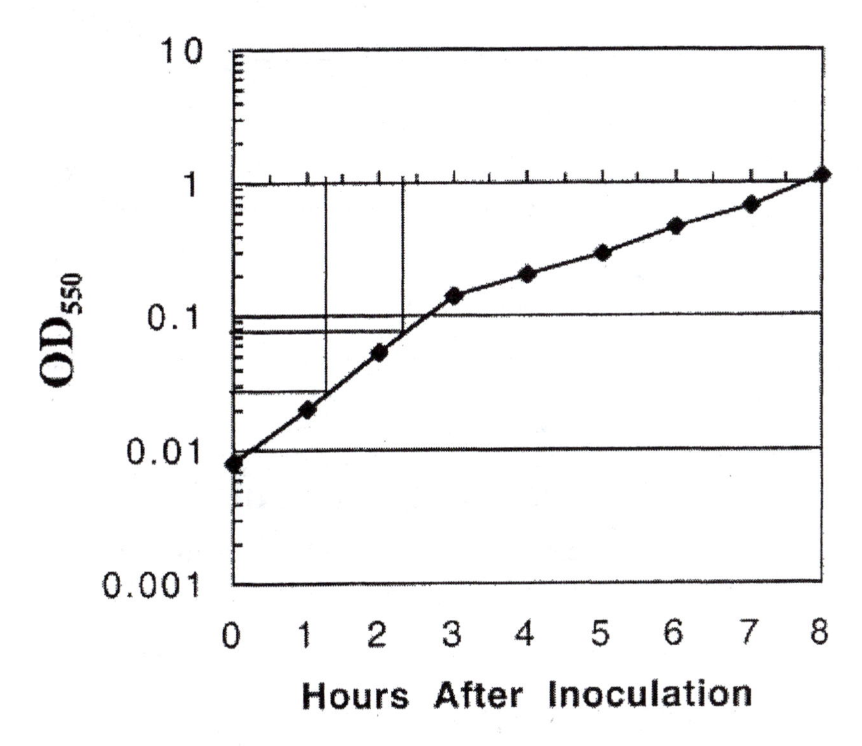

Figure 3.5 Determination of generation time for Problem 3.12.

The difference between the points where these lines meet the *x* axis is approximately 0.8 hours. Since 0.8 hours is equivalent to 48 minutes (0.8 hr × 60 min/hr = 48 min), the generation time is approximately 48 minutes.

Plotting Cell Concentration vs. Time on a Semilog Graph

When cell concentration is plotted against time (the number of hours following inoculation) on a semilog graph, the plot in Figure 3.6 is generated. [Since cell concentration ranges with exponent values between 7 and 9 (covering three exponent values), three-cycle semilog graph paper is be used.]

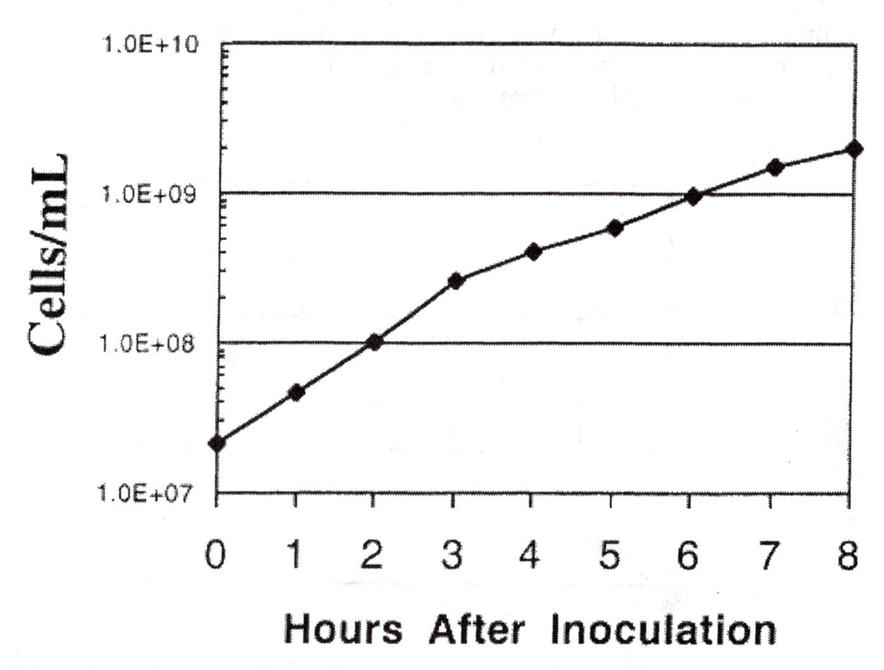

Figure 3.6 Semilog plot of cell concentration vs. incubation time for sample data. The notation 1.0E+07 on the *y* axis is shorthand for 1.0×10^7. The *E* in this notation stands for "exponent."

Problem 3.13 Assuming that a culture of cells is grown again under identical conditions to those used to generate the sample data in Table 3.1, at what point will the culture have a concentration of 2×10^8 cells/mL?

Solution 3.13 Drawing a horizontal line from the 2×10^8 cells/mL point on the *y* axis to the plotted curve and then vertically down from that position to the *x* axis leads to a value of 2.7 hours (see Figure 3.7). Since 0.7 hours is equivalent to 42 minutes (0.7 hr × 60 min/hr = 42 min), the culture will reach a concentration of 2×10^8 cells/mL in 2 hours 42 minutes.

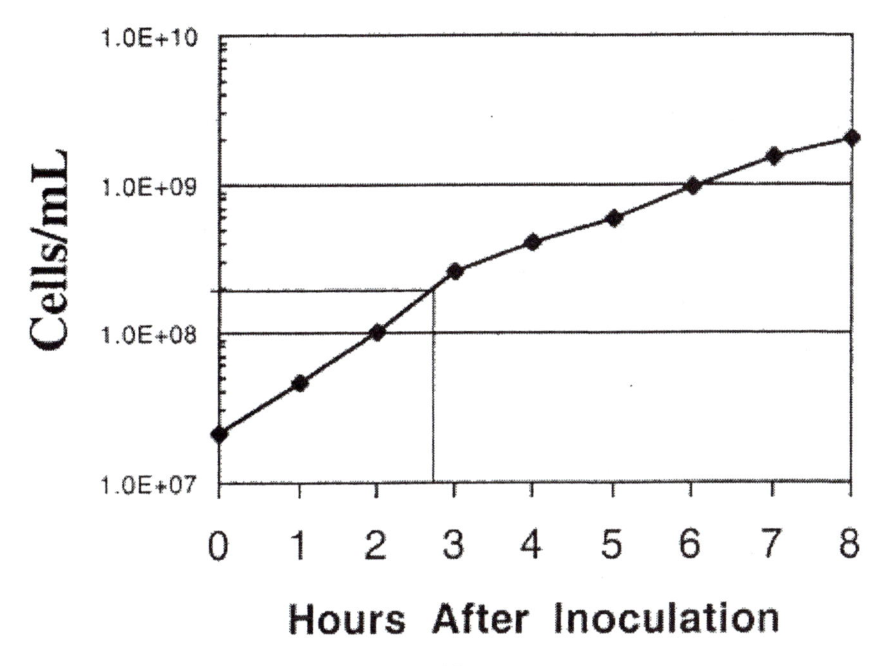

Figure 3.7 Plot for Problem 3.13.

Determining Generation Time Directly from a Semilog Plot of Cell Concentration vs. Time

The generation time for a particular bacterial strain in culture can be read directly from the semilog plot of cell concentration vs. time in a similar manner to that described in Problem 3.13.

Problem 3.14 Using a plot of OD_{550} vs. time, determine the generation time for those cells represented by the sample data in Table 3.1.

Solution 3.14 Choose a point on the *y* axis (cells/mL) in the region corresponding to the exponential growth phase. For example, a cell concentration value of 1×10^8 cells/mL can be chosen. Twice this value is 2×10^8 cells/mL. Draw horizontal lines from these two positions on the *y* axis to the plotted curve (see Figure 3.8). From those intersection points on the curve, draw vertical lines to the *x* axis. The difference between the positions where these lines meet the *x* axis is approximately 0.7 hours. Since 0.7 hours is equivalent to 42 minutes (0.7 hr $\times$ 60 min/hr = 42 min), the generation time is approximately 42 minutes.

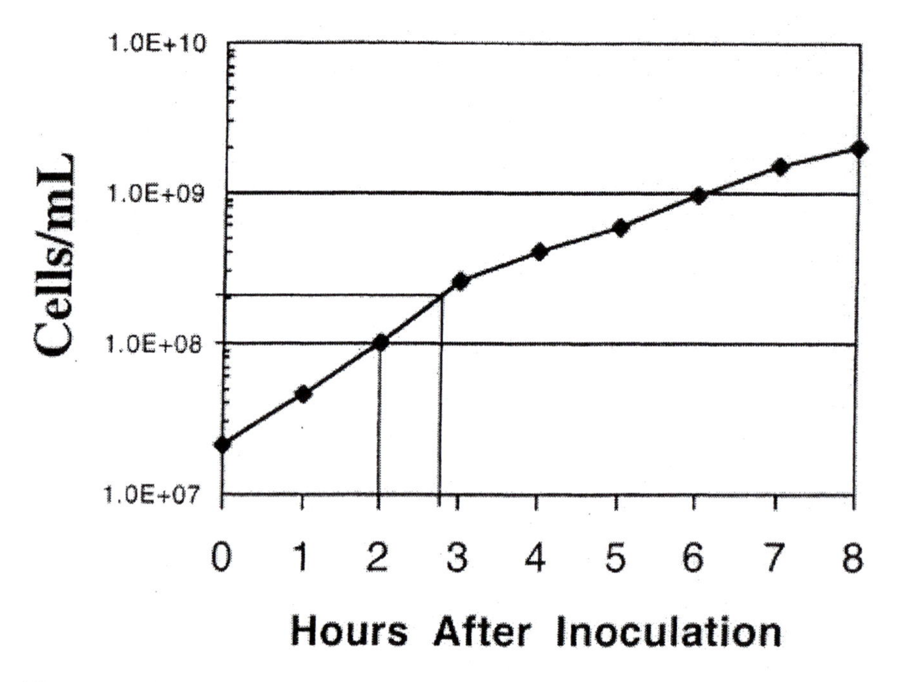

Figure 3.8 Plot for Problem 3.14.

Plotting Cell Density vs. OD_{550} on a Semilog Graph

Plotting cell density vs. OD_{550} on a semilog graph allows the experimenter to quickly estimate cell concentration for any OD_{550} value. For the sample data in Table 3.1, this plot gives the curve shown in Figure 3.9.

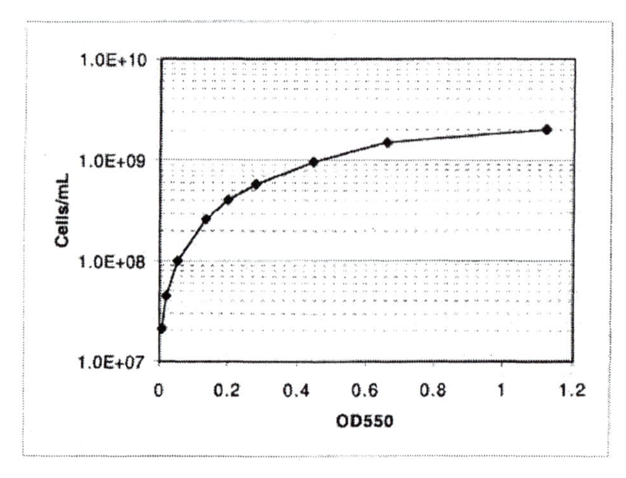

Figure 3.9 Semilog plot of cell concentration vs. OD_{550} for sample data.

Problem 3.15 A culture is grown under identical conditions to those used to generate the sample data. If a sample is withdrawn and its OD_{550} is determined to be 0.31, what is the approximate cell concentration?

Solution 3.15 To determine the cell concentration corresponding to an OD_{550} value of 0.31, draw a vertical line from the 0.31 position on the *x* axis up to the plotted curve (Figure 3.10). Draw a horizontal line from this position on the curve to the *y* axis.

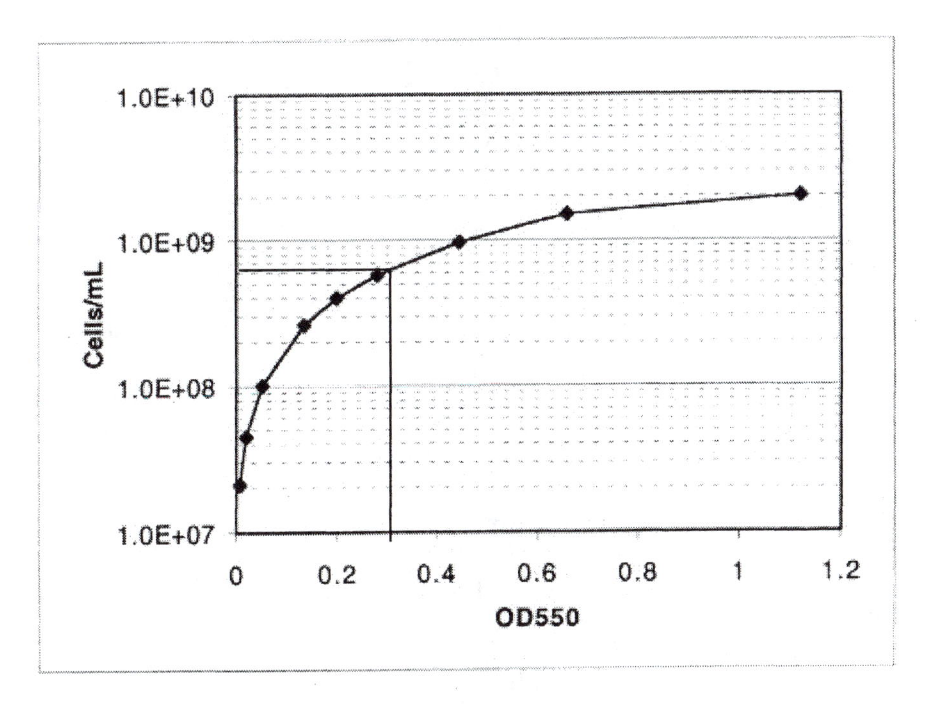

Figure 3.10 Plot for Problem 3.15.

This point on the *y* axis corresponds to 6×10^8 cells/mL. Therefore, a culture having an OD_{550} of 0.31 will have a cell concentration of 6×10^8 cells/mL.

The Fluctuation Test

Bacteriophages (or **phages**) are viruses that infect bacteria. A number of different types of bacteriophages infect the bacterium *Escherichia coli.* When infecting their host, many of the best characterized bacteriophages undertake a lytic course of infection, in which the phage particle attaches to the host's cell wall and subsequently injects its genome into the cell's cytoplasm, where it replicates and expresses the genes necessary for the construction of progeny phage. During the infection process, progeny phage accumulate inside the cell. Ultimately, the cell bursts open (lyses), releasing the new phage particles into the environment, where they can infect other cells.

If bacteriophage are grown in liquid culture, infection and phage multiplication leads to a decrease in the culture's turbidity as more and more cells succumb to infection and lyse. If bacteriophage are plated with cells on a petri plate, infection of a single cell by a single phage eventually leads to a circular area of lysis (called a **plaque**) within the bacterial lawn.

The bacteriophage T1 follows the infection pathway just described. T1 initiates infection by first binding to a protein complex, called the T1-phage receptor site, on the host's cell wall. Once attachment is accomplished, infection leads to the eventual destruction of the attacked cell. Not all cells in an *E. coli* population, however, are susceptible to infection. Within a population of cells, there is a chance that some members of that population carry a mutation that alters the T1 receptor. Such cells are resistant to T1 infection. A T1-resistant cell passes this characteristic on to daughter cells in a stable manner.

In the early 1940s, as more and more mutations were being characterized in both prokaryotic and eukaryotic systems, an essential question remained unanswered: What is the origin of such genetic variants? Two possibilities were considered: Either mutations are induced as a response to the environment, or they arise spontaneously; preexisting in a population, the environment merely selects for their survival.

In 1943, Max Delbrück and Salvador Luria designed an experiment using T1-resistant mutants of *E. coli* to determine which hypothesis was correct. The experiment entailed looking at the number of T1-resistant cells in a parallel series of identical cultures. By the first hypothesis, that of induced mutation, T1-resistant variants arise only upon exposure to T1 bacteriophage. All cells have an equal (but small) probability of acquiring T1 resistance following exposure to T1. Once acquired, the adaptation is passed on to their descendants. This hypothesis predicts that, no matter the stage of cell growth, the fraction of T1-resistant cells should fluctuate neither between a series of cultures nor between a series of samples taken from the same bulk culture.

By the second hypothesis, that of spontaneous mutation, the acquisition of T1 resistance may occur at any time during growth of a culture prior to T1 exposure. The number of T1-resistant cells in any particular population will depend upon whether the mutation occurred early or late in the growth of that culture. This theory predicts larger fluctuations in the number of T1-resistant cells between cultures in a parallel series than from a series of samples taken from the same bulk culture.

Fluctuation Test Example

A series of 10 tubes containing 0.2 mL of tryptone broth and a single 125-mL containing 10 mL of tryptone broth are each inoculated with *E. coli* to a concentration of 500 cells/mL. Cells are incubated at 37°C until the cultures reach stationary phase of growth. One-tenth-mL samples are taken from each of the small cultures and 10 0.1-mL aliquots are taken from the 10-mL bulk culture. These samples are then spread on agar plates saturated with T1 phage. The plates are incubated overnight and the number of colonies on each plate is counted. The results shown in Table 3.4 are obtained.

T1-Resistant Colonies from the Small Cultures	T1-Resistant Colonies from the Bulk Culture
1	24
0	20
0	16
5	25
124	18
0	21
0	18
0	17
71	22
0	19

Table 3.4 Sample data for fluctuation test problems.

Interpretation of Results. By visual inspection of the data, there is an obvious difference between the distribution of T1-resistant colonies derived from the different cultures. As predicted by the theory of spontaneous mutation, there are large differences in the numbers of T1-resistant colonies derived from the parallel series of small cultures, while the number of T1-resistant colonies per plate arising from samples of the bulk culture shows little variation. To examine these results mathematically, however, it is first necessary to determine the mean value for each set of cultures. A **mean** (also called an **average)** represents a value within the range of an entire set of numbers. It falls between the two extreme values of a number set and serves as a representative value for any set of measurements. It is calculated as the sum of the measures in a distribution divided by the number of measures. The mean value is represented by the symbol x with a line over it:

$$\text{mean} = \bar{x}$$

Problem 3.16 What are the mean values for each data set?

Solution 3.16 For the parallel series of small cultures, the sum of T1-resistant colonies is

$$1 + 0 + 0 + 5 + 124 + 0 + 0 + 0 + 71 + 0 = 201$$

The mean value is then calculated as the sum divided by the number of measures taken:

$$\frac{201 \text{ T1-resistant colonies}}{10 \text{ samples}} = 20.1 \text{ T1-resistant colonies/sample}$$

For the samples taken from the 10-mL bulk culture, the sum of T1-resistant colonies is

$$24 + 20 + 16 + 25 + 18 + 21 + 18 + 17 + 22 + 19 = 200$$

The mean value is then calculated as this sum divided by the number of samples taken:

$$\frac{200 \text{ T1-resistant colonies}}{10 \text{ samples}} = 20.0 \text{ T1-resistant colonies/sample}$$

Variance

The spontaneous mutation hypothesis predicts a large fluctuation around the mean value for the samples taken from the individual cultures. Fluctuation within a data set around the set's average value is referred to as the **variance**. Although the mean values of T1-resistant colonies between the two sets of data should be quite similar, as might be predicted by either hypothesis, to satisfy the spontaneous mutation hypothesis, the variance between the two sets of samples should be large.

Variance is calculated by the formula

$$\text{variance} = \frac{\sum (x - \bar{x})^2}{n - 1}$$

By this formula, the term $(x - \bar{x})^2$ requires that the mean value for each data set be subtracted from each individual value in that data set. The remainder of each subtraction is then squared. (**Squaring** a number means to multiply that number or quantity by itself. It is denoted by the exponent 2. It can be obtained on a calculator by entering the number to be squared and then pressing the x^2 key.) The symbol Σ (the Greek letter **sigma**) demands that all terms in a series be added. In this case, the series is the $(x - \bar{x})^2$ values obtained for each data point. The number n is equal to the number of measurements.

Problem 3.17 What is the variance for each data set?

Solution 3.17 Calculating the sum of squares for the small-cultures data set is demonstrated in Table 3.5.

T1-Resistant Colonies	$x - \bar{x}$	$(x - \bar{x})^2$
1	1 – 20 = –19	$(-19)^2 = 361$
0	0 – 20 = –20	$(-20)^2 = 400$
0	0 – 20 = –20	$(-20)^2 = 400$
5	5 – 20 = –15	$(-15)^2 = 225$
124	124 – 20 = 104	$(104)^2 = 10{,}816$
0	0 – 20 = –20	$(-20)^2 = 400$
0	0 – 20 = –20	$(-20)^2 = 400$
0	0 – 20 = –20	$(-20)^2 = 400$
71	71 – 20 = 51	$(51)^2 = 2601$
0	0 – 20 = –20	$(-20)^2 = 400$
		$\Sigma(x - \bar{x})^2 = 16{,}403$

Table 3.5 Calculating the sum of squares of the mean subtracted from each value in the small-cultures data for Problem 3.17.

The sum of the values in the $(x - \bar{x})^2$ column = $\Sigma(x - \bar{x})^2 = 16{,}403$.

Since there are 10 samples in this data set, $n = 10$.

Placing these values into the formula for calculating variance yields

$$\text{Variance} = \frac{\sum(x-\bar{x})^2}{n-1} = \frac{16{,}403}{10-1} = \frac{16{,}403}{9} = 1823$$

Calculating the sum of squares for the bulk-culture data set is demonstrated in Table 3.6.

T1-Resistant Colonies	$x-\bar{x}$	$(x-\bar{x})^2$
24	24 – 20 = 4	$(4)^2 = 16$
20	20 – 20 = 0	$(0)^2 = 0$
16	16 – 20 = –4	$(-4)^2 = 16$
25	25 – 20 = 5	$(5)^2 = 25$
18	18 – 20 = –2	$(-2)^2 = 4$
21	21 – 20 = 1	$(1)^2 = 1$
18	18 – 20 = –2	$(-2)^2 = 4$
17	17 – 20 = –3	$(-3)^2 = 9$
22	22 – 20 = 2	$(2)^2 = 4$
19	19 – 20 = –1	$(-1)^2 = 1$
		$\Sigma(x-\bar{x})^2 = 80$

Table 3.6 Calculating the sum of squares of the mean subtracted from each value in the bulk-culture data for Problem 3.17.

The sum of the values in the $(x-\bar{x})^2$ column = $\Sigma(x-\bar{x})^2 = 80$.

Since there are 10 samples in this data set, $n = 10$.

Placing these values into the formula for calculating variance yields

$$\text{variance} = \frac{\sum(x-\bar{x})^2}{n-1} = \frac{80}{10-1} = \frac{80}{9} = 8.9$$

Unlike the mean values for the two sets of samples, the variance values are drastically different and support the spontaneous mutation hypothesis.

Measuring Mutation Rate

Luria and Delbrück demonstrated that data from a fluctuation test can be used, by two different methods, to calculate a spontaneous mutation rate. One of these methods utilizes the law of probabilities as described by a Poisson distribution. The other utilizes a graphical approach, taking into account the frequency distribution of mutant cells in the parallel series of small cultures.

The Poisson Distribution

The **Poisson distribution** is used to describe the distribution of rare events in a large population. For example, at any particular time, there is a certain probability that a particular cell within a large population of cells will acquire a mutation. Mutation acquisition is a rare event. If the large population of cells is divided into smaller cultures, as is done in the fluctuation test, the Poisson distribution can be used to determine the probability that any particular small culture will contain a mutated cell.

Calculating a Poisson distribution probability requires the use of the number ***e***, described in the following box.

The Number *e*

In molecular biology, statistics, physics, and engineering, most calculations employing the use of logarithms are in one of two bases, either base 10 or base *e*. The number *e* is the base of the **natural logarithms**, designated as **ln**. For example, ln 2 is equivalent to $\log_e 2$. The value of *e* is roughly equal to 2.7182818. *e* is called an irrational number because its decimal representation neither terminates nor repeats. In that regard, it is like the number pi (π, the ratio of the circumference of a circle to its diameter). In fact, pi and *e* are related by the expression $e^{i\pi} = -1$, where *i* is equal to the square root of –1.

Many calculators have an **ln** key for finding natural logarithms. Many calculators also have an e^x key used to find the antilogarithm base *e*.

The Poisson distribution is written mathematically as

$$P = \frac{e^{-m} m^r}{r!}$$

where *P* is the fraction of samples that will contain *r* objects each, if an average of *m* objects per sample is distributed at random over the collection of samples. (The *m* component is sometimes referred to as the **expectation**.) The component *e* is the base of the **natural** system of logarithms (see box). The exclamation mark, !, is the symbol for factorial. The **factorial** of a number is the product of the specified number and each positive integer less than itself down to and including 1. For example, 5! (read "5 factorial") is equal to $5 \times 4 \times 3 \times 2 \times 1 = 120$.

0! is equal to **1**

Calculating Mutation Rate by Using the Poisson Distribution
To use the fluctuation test to determine mutation rate via the Poisson distribution, the small cultures must be prepared with a sufficiently small inocula such that, following incubation, some of them will have no mutants at all. In addition, it is important to assay the total number of cells per culture at the time the culture is spread on selective plates.

Mutation rate is calculated using the fraction of small parallel cultures that contain no mutants. To solve the Poisson distribution for the zero case, the equation is

$$P_0 = \frac{e^{-m}m^0}{0!}$$

For our example, 6 of the 10 small cultures contained no T1-resistant bacteria. P_0, therefore, the fraction of cultures containing no mutants, is 6/10 = 0.6. The Poisson distribution relationship can now be written as

$$\frac{6}{10} = 0.6 = \frac{e^{-m}m^0}{0!}$$

The component m^0 is equivalent to 1. (Any number raised to the exponent 0 is equal to 1.) Likewise, 0! is equal to 1. The equation then reduces to

$$0.6 = e^{-m}$$

We can solve for m using the following relationship:

$\boldsymbol{x = e^y}$ is equivalent to $\boldsymbol{y = \ln x}$
and
$\boldsymbol{x = e^{-y}}$ is equivalent to $\boldsymbol{y = -\ln x}$.

$$m = -\ln 0.6$$

To determine the natural log of 0.6 on the calculator, press **.**, **6**, and **ln**. This gives a value, when rounded, of –0.51.

$$m = -(-0.51)$$

The negative of a negative number is a positive value. Therefore,

$$m = 0.51$$

That is, m, the average number of mutant bacteria in a series of cultures representing an entire population, is equivalent to 0.51 mutants per small culture.

To determine a mutation rate as mutations/cell/cell division, the number of mutants per small culture (in our example, 0.51) must be divided by the number of bacteria per cell division. The following relationship can be used to calculate that value.

Calculating the Increase in Bacteria from Generation to Generation

The average number of bacteria in a subsequent generation is calculated by dividing the number of bacteria at the beginning of the current generation by the natural logarithm of 2.

Therefore, if there were 2.6×10^8 bacteria per small culture (as assayed at the time the cultures were spread on the selective plates containing T1 bacteriophage), then the number of bacteria per cell division is equal to

$$2.6 \times 10^8/\ln 2 = 2.6 \times 10^8/0.69$$

$$= 3.8 \times 10^8$$

The mutation rate is then calculated as

$$0.51/(3.8 \times 10^8) = 1.3 \times 10^{-9} \text{ mutations/cell/cell division.}$$

Using a Graphical Approach to Calculate Mutation Rate from Fluctuation Test Data

Using the Poisson distribution method to calculate mutation rate does not make use of all the information available from a fluctuation test; it makes no use of the frequency distribution of cultures that do contain mutant (T1-resistant) cells. A second approach developed by Luria and Delbrück relies on the assumption that, once a population of cells reaches a certain size, at least one cell will acquire a mutation. In the subsequent generation, as the number of cells doubles (and assuming a constant mutation frequency), two cells should newly acquire mutations. In the next generation, following another doubling, four cells should newly acquire mutations. The fraction of cells acquiring mutations is assumed to stay constant over time. As a simplified example, if a single cell in a population of four cells acquires a mutation, then one-fourth of the cells have acquired a mutation. In the next generation, after the cells have doubled, two out of the eight cells (one-fourth) would be expected to acquire mutations. During the next doubling, four out of the sixteen daughter cells (one-fourth) would be expected to acquire mutations. Although the mutation rate should remain constant, the overall proportion of mutant cells in the population will increaseas those cells acquiring mutations pass that new trait along to daughter cells in subsequent generations.

The equation developed by Luria and Delbrück to account for this scenario and to determine the time at which there is likely to be a single mutant somewhere in the series of small cultures is

$$r = aN_t \ln(aN_tC)$$

where

r = average number of mutants per small culture
C = number of small cultures used in the fluctuation test
a = mutation rate
N_t = total number of cells in each small culture

In the example provided in this text,

r = 20
C = 10
$N_t = 2.6 \times 10^8$

The value of aN_t in the equation can be solved by interpolation through the systematic substitution of arbitrary values of a and values of N_t. A graph of r vs. aN_t when C is equal to 10 is then used to solve for the actual mutation rate, a. Mutation rates (a) typically vary from 1×10^{-8} to 1×10^{-12}. The number of bacteria (N_t) in the small cultures, depending on the strain and the media used, may vary from 5×10^7 to 5×10^{10}. To prepare a graph of r vs. aN_t, various values of a are multiplied by various values of N_t, shown in Tables 3.7 – 3.11. Once aN_t is determined, r can be solved via the mutation rate equation. For example, when $a = 5 \times 10^{-7}$ and $N_t = 5 \times 10^7$, then $aN_t = 25$. Using the mutation rate equation,

$$r = aN_t \ln(aN_tC)$$

Substituting our known values yields

$$r = 25 \ \ln(25 \times 10) = 25 \ \ln 250 = (25)(5.5) = 138$$

Tables 3.7 – 3.11 show values of r obtained when various values of a and N_t are used.

a	N_t	aN_t	r
5×10^{-7}	5×10^{7}	25.0	138.0
5×10^{-8}	5×10^{7}	2.5	8.0
2×10^{-8}	5×10^{7}	1.0	2.3
1×10^{-8}	5×10^{7}	0.5	0.8
5×10^{-9}	5×10^{7}	0.25	0.23

Table 3.7 r for various values of a when N_t is 5×10^{7} and $C = 10$.

a	N_t	aN_t	r
5×10^{-7}	1×10^{8}	50.0	310.0
5×10^{-8}	1×10^{8}	5.0	19.6
2×10^{-8}	1×10^{8}	2.0	6.0
1×10^{-8}	1×10^{8}	1.0	2.3
5×10^{-9}	1×10^{8}	0.5	0.8
1×10^{-9}	1×10^{8}	0.1	0.0

Table 3.8 r for various values of a when N_t is 1×10^{8} and $C = 10$.

a	N_t	aN_t	r
5×10^{-7}	2×10^{8}	100.0	691.0
5×10^{-8}	2×10^{8}	10.0	46.1
2×10^{-8}	2×10^{8}	4.0	14.8
1×10^{-8}	2×10^{8}	2.0	6.0
5×10^{-9}	2×10^{8}	1.0	2.3
1×10^{-9}	2×10^{8}	0.2	0.14
5×10^{-10}	2×10^{8}	0.1	0.0

Table 3.9 r for various values of a when N_t is 2×10^{8} and $C = 10$.

a	N_t	aN_t	r
5×10^{-7}	5×10^{8}	250.0	1956.0
5×10^{-8}	5×10^{8}	25.0	138.0
2×10^{-8}	5×10^{8}	10.0	46.1
1×10^{-8}	5×10^{8}	5.0	19.6
5×10^{-9}	5×10^{8}	2.5	8.0
1×10^{-9}	5×10^{8}	0.5	0.8
5×10^{-10}	5×10^{8}	0.25	0.23

Table 3.10 r for various values of a when N_t is 5×10^{8} and $C = 10$.

a	N_t	aN_t	r
5×10^{-7}	1×10^{9}	500.0	4259.0
5×10^{-8}	1×10^{9}	50.0	311.0
2×10^{-8}	1×10^{9}	20.0	106.0
1×10^{-8}	1×10^{9}	10.0	46.1
5×10^{-9}	1×10^{9}	5.0	19.6
1×10^{-9}	1×10^{9}	1.0	2.3
5×10^{-10}	1×10^{9}	0.5	0.8
1×10^{-10}	1×10^{9}	0.1	0.0

Table 3.11 r for various values of a when N_t is 1×10^9 and $C = 10$.

Plotting r vs. aN_t for $C = 10$ gives the graph presented in Figure 3.11.

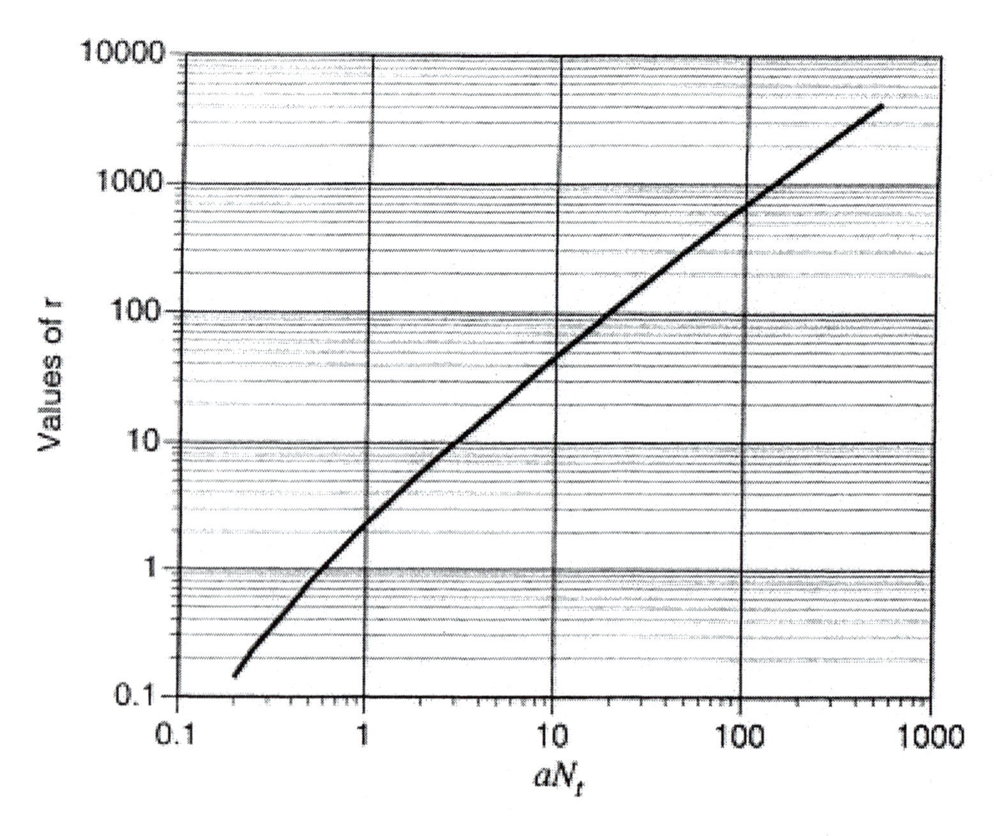

Figure 3.11 Graph of r plotted against aN_t for $C = 10$.

From Figure 3.12, it can be seen that when $r = 20$, $aN_t = 5$.

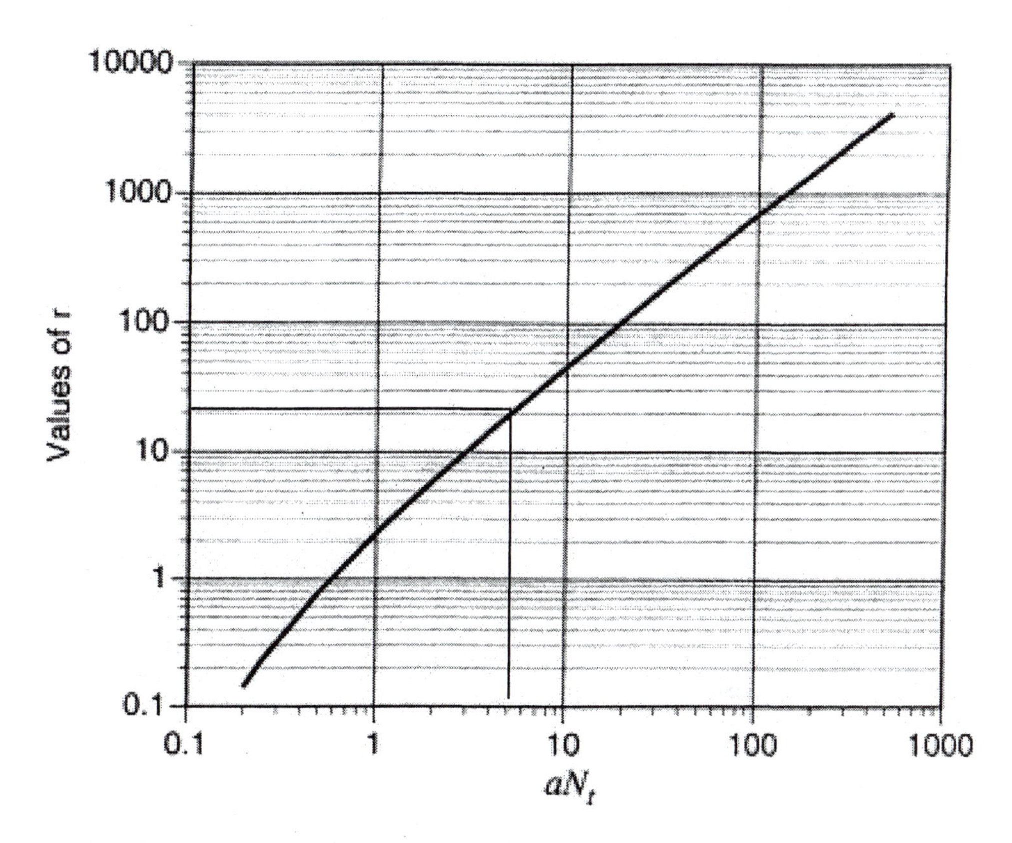

Figure 3.12 Determining aN_t when $r = 20$.

Therefore, using the equation $r = aN_t \ln(aN_tC)$, where $r = 20$, $N_t = 2.6\times10^8$, $C = 10$, and aN_t is equivalent to 5, we have

$$20 = a(2.6\times10^8)\ln(5 \text{ x } 10)$$

$$20 = a(2.6\times10^8)\ln(50)$$

$$20 = a(2.6\times10^8)(3.9)$$

Solving for a, the mutation rate, gives the following solution:

$$a = \frac{20}{\left(2.6\times10^8\right)(3.9)} = \frac{20}{1.014\times10^9} = 2\times10^{-8}$$

This mutation rate (2×10^{-8}) is slightly higher than that calculated by the Poisson distribution method. This is because this method is sensitive to the appearance of "jackpot" tubes, in which a mutation occurred very early following inoculation such that a large number of mutant cells appears in the final population. Jackpot tubes result in excessively large values of r and an overestimate of the actual number of mutant cells arising during a defined period of time.

Mutation Rate Determined by Plate Spreading

T1-resistant mutants can be detected by spreading cells on an agar plate, spraying an aerosol of T1 bacteriophage onto the plate, and then allowing time for the appearance of T1-resistant colonies. This approach to mutation detection can be used, with slight modifications, to determine a mutation frequency. For example, the following protocol could be followed to arrive at such a number.

1. Begin with a culture of *E. coli* cells at a concentration of 2×10^5 cells/mL.

2. With a sterile plate spreader, spread 0.1 mL (2×10^4 cells) onto each of 24 tryptone plates and onto each of 20 plates previously spread with T1 bacteriophage. This latter series of plates is used to assay for the number of T1-resistant cells in the initial culture.

3. Allow the spread plates to incubate for 6 hours. [Assuming a generation time of 40 minutes, this allows for 9 doublings (360 min ÷ 40 min = 9).]

4. At 6 hours, spray an aerosol of T1 bacteriophage onto 20 of the 24 plates not previously treated with T1, and incubate these sprayed plates for 12 hours to allow the growth of resistant colonies. From the remaining four untreated plates, collect and assay the number of cells.

This experiment might give the following results.

Initial number of T1-resistant cells (as determined by the number of colonies appearing on the 20 plates previously spread with T1 phage) = 0.

Initial number of bacteria =

$$\frac{2\times10^5 \text{ cells}}{\text{mL}}\times\frac{0.1 \text{ mL}}{\text{plate}}\times 20 \text{ plates} = 4\times10^5 \text{ cells}$$

After 6 hours of growth, the four untreated plates are assayed and found to have 1×10^{7} cells/plate. Therefore, on the 20 plates sprayed with a T1 aerosol, there are a total of 2×10^{8} cells:

$$20 \text{ plates} \times \frac{1\times10^{7} \text{ cells}}{\text{plate}} = 2\times10^{8} \text{ total cells}$$

Following incubation of the 20 sprayed plates, five colonies arise.

The mutation rate can be calculated using the following formula:

$$\textbf{mutation rate} = \frac{\textbf{change in number of resistant colonies/change in total number of cells}}{\textbf{ln 2}}$$

In the example provided, the change in the number of T1-resistant colonies is

$$5-0=5$$

The change in the number of total cells is

$$2\times10^{8} \text{ cells} - 4\times10^{5} \text{ cells} = 2\times10^{8} \text{ cells}$$

(Subtracting 4×10^{5} from 2×10^{8} yields 2×10^{8} when significant figures are considered.)

Using the formula given above for mutation rate yields

$$\text{mutation rate} = \frac{5/2\times10^{8}}{\ln 2} = \frac{2.5\times10^{-8}}{0.693} = 3.6\times10^{-8}$$

Therefore, the mutation rate is 3.6×10^{-8} mutations/bacterium/cell division.

Measuring Cell Concentration on a Hemocytometer

A hemocytometer is a microscope slide engraved with a 1-mm×1-mm grid. A coverslip rests on supports 0.1-mm above the grid such that 0.1-mm^{3} chambers are formed by the 1-mm×1-mm squares of the grid and the 0.1-mm space between the slide and the coverslip. A drop of cell suspension is drawn into the chamber below the coverslip, and the slide is examined under a microscope. The observer then counts the number of cells in several of the 1-mm×1-mm squares. (For best accuracy, a total of over 200 cells should be counted.) Cell count is then converted to a concentration, as demonstrated in the following problem.

Problem 3.18 A suspension of hybridoma cells is diluted 1 mL/10 mL and an aliquot of the dilution is counted on a hemocytometer. In a total of 10 grid squares, 320 cells are counted. What is the cell concentration of the cell suspension?

Solution 3.18 To determine cell concentration, the conversion of area (mm^3) to volume must be used. One milliliter of liquid is equal to 1 cm^3 (one cubic centimeter). One cm^3 = 1000 mm^3. There are 10 0.1-mm^3 per cubic millimeter. These conversion factors and the dilution factor will be used to determine cell concentration, as shown in the following equation:

$$\frac{320 \text{ cells}}{10 \ 0.1-\text{mm}^3 \text{ squares}} \times \frac{10 \ 0.1-\text{mm}^3 \text{ squares}}{\text{mm}^3} \times \frac{1000 \text{ mm}^3}{\text{cm}^3} \times \frac{1 \text{ cm}^3}{\text{mL}} \times \frac{10 \text{ mL}}{1 \text{ mL}} = \frac{3.2 \times 10^6 \text{ cells}}{\text{mL}}$$

Therefore, the cell suspension has a concentration of 3.2×10^6 cells/mL.

Reference

Luria, S.E., and M. Delbrück (1943). Mutations of bacteria from virus sensitivity to virus resistance. *Genetics* 28:491–511.

Working with Bacteriophage

4

Introduction

A **bacteriophage (phage)** is a virus that infects bacteria. It is little more than nucleic acid surrounded by a protein coat. To infect a cell, a bacteriophage attaches to a receptor site on the bacteria's cell wall. Upon attachment, the phage injects its DNA into the cell's cytoplasm, where it is replicated. Phage genes are expressed for the production and assembly of coat proteins that encapsulate the replicated phage DNA. When a critical number of virus particles has been assembled, the host cell lyses and the newly made phage are released into the environment, where they can infect new host cells. Their simple requirements for propagation, their short generation time, and their relatively simple genetic structure have made bacteriophages ideal subjects of study for elucidating the basic mechanisms of transcription, DNA replication, and gene expression. A number of bacteriophages have been extensively characterized. Several, such as the bacteriophage lambda and the bacteriophage M13, have been genetically engineered to serve as vectors for the cloning of exogenous genetic material.

Multiplicity of Infection

An experiment with bacteriophage typically begins with an initial period during which the virus is allowed to adsorb to the host cells. It is important to know the ratio of the number of bacteriophages to the number of cells at this stage of the infection process. Too many bacteriophages attaching to an individual cell can cause cell lysis, even before the infection process can yield progeny virus particles ("lysis from without"). If too few bacteriophage are used for the infection, it may be difficult to detect or measure the response being tested. The bacteriophage to cell ratio is called the **multiplicity of infection (m.o.i.)**.

Problem 4.1 A 0.1-mL aliquot of a bacteriophage stock having a concentration of 4×10^9 phage/mL is added to 0.5 mL of *E. coli* cells having a concentration of 2×10^8 cells/mL. What is the m.o.i.?

Solution 4.1 First, calculate the total number of bacteriophage and the total number of bacteria.

Total number of bacteriophage:

$$0.1\ \text{mL} \times \frac{4 \times 10^{9}\ \text{phage}}{\text{mL}} = 4 \times 10^{8}\ \text{bacteriophage}$$

Total number of cells:

$$0.5\ \text{mL} \times \frac{2 \times 10^{8}\ \text{cells}}{\text{mL}} = 1 \times 10^{8}\ \text{cells}$$

The m.o.i. is then calculated as bacteriophage per cell:

$$\text{m.o.i.} = \frac{4 \times 10^{8}\ \text{phage}}{1 \times 10^{8}\ \text{cells}} = 4\ \text{phage/cell}$$

Therefore, the m.o.i. is 4 phage/cell.

Problem 4.2 A 0.25-mL aliquote of an *E. coli* culture having a concentration of 8×10^{8} cells/mL is placed into a tube. What volume of a bacteriophage stock having a concentration of 2×10^{9} phage/mL should be added to the cell sample to give an m.o.i. of 0.5?

Solution 4.2 First, calculate the number of cells in the 0.25-mL aliquot:

$$0.25\ \text{mL} \times \frac{8 \times 10^{8}\ \text{cells}}{\text{mL}} = 2 \times 10^{8}\ \text{cells}$$

Next, calculate how many bacteriophage particles are required for an m.o.i. of 0.5 when 2×10^{8} cells are used.

$$\frac{x\ \text{phage}}{2 \times 10^{8}\ \text{cells}} = \frac{0.5\ \text{phage}}{\text{cell}}$$

Solve for x:

$$x = \left(\frac{0.5\ \text{phage}}{\text{cell}}\right)\left(2 \times 10^{8}\ \text{cells}\right) = 1 \times 10^{8}\ \text{phage}$$

Finally, calculate the volume of the bacteriophage stock that will contain 1×10^{8} phage:

$$\frac{2 \times 10^{9}\ \text{phage}}{\text{mL}} \times x\ \text{mL} = 1 \times 10^{8}\ \text{phage}$$

Solving for x yields

$$x = \frac{1 \times 10^8 \text{ phage}}{2 \times 10^9 \text{ phage/mL}} = 0.05 \text{ mL}$$

Therefore, 0.05 mL of the phage stock added to the aliquot of cells will give an m.o.i. of 0.5.

Probabilities and Multiplicity of Infection

In the previous chapter, probability was used to estimate the number of mutant cells in a culture of bacteria. Probability can also be used to examine infection at the level of the individual cell and to estimate the distribution of infected cells in culture. These are demonstrated in the following problems.

Problem 4.3 A culture of bacteria is infected with bacteriophage at an m.o.i. of 0.2. What is the probability that any one cell will be infected by two phage?

Solution 4.3 When the multiplicity of infection is less than 1, the math used to calculate the chance that any particular cell will be infected is similar to that used to predict a coin toss. The probability that any one cell will be infected by a single virus is equal to the m.o.i. (in this case 0.2). This also means that 20% of the cells will be infected ($100 \times 0.2 = 20\%$) or that each cell has a 20% chance of being infected. This value can further be expressed as a "1 in" number by taking its reciprocal:

$$\frac{1}{0.2} = 5$$

Therefore, 1 in every 5 cells will be infected.

If it is assumed that attachment of one phage does not influence the attachment of other phage, then the attachment of a second phage will have the same probability as the attachment of the first. The probabilities of both events can be multiplied.

The probability that a cell will be infected by two phage, therefore, is the product of the probabilities of each independent event:

$$0.2 \times 0.2 = 0.04$$

This can be expressed in several ways:

a) 4% of the cells will have two phage particles attached ($0.04 \times 100 = 4\%$), or
b) a cell has a 4% chance of being infected by two phage particles, or

c) 1 in 25 cells will be infected by two phage particles (the reciprocal of 0.04 = 1/0.04 = 25).

Problem 4.4 A culture of cells is infected by bacteriophage lambda at an m.o.i. of 5. What is the probability that a particular cell will not be infected during the phage adsorption period?

Solution 4.4 In the previous chapter, the Poisson distribution was used to determine the number of mutants that might be expected in a population of cells. This distribution is represented by the equation

$$P_r = \frac{e^{-m} m^r}{r!}$$

where *P* is the probability, *r* is the number of successes (in this example, a cell with zero attached phage is a "success"), *m* is the average number of phage/cell (the m.o.i.; 5 in this example), and *e* is the base of the natural logarithms.

For the zero case, the equation becomes

$$P_0 = \frac{e^{-5} 5^0}{0!}$$

In solving for the zero case, the following two properties are encountered.

Any number raised to the zero power is equal to 1.

The exclamation symbol (!) designates the factorial of a number, which is the product of all integers from that number down to 1. For example, 4! (read "4 factorial" or "factorial 4") is equal to $4 \times 3 \times 2 \times 1 = 24$. 0! is equal to 1.

The Poisson distribution for the zero case becomes

$$P_0 = \frac{e^{-5}(1)}{1} = e^{-5}$$

e^{-5} is equal to 0.0067 (on the calculator: **5**, **+/–**, then e^x).

Therefore, at an m.o.i. of 5, the probability that a cell will not be infected is 0.0067. This is equivalent to saying that 0.67% of the culture will be uninfected (100 × 0.0067 = 0.67%) or that 1 in 149 cells will be uninfected (1/0.0067 = 149).

Problem 4.5 A culture of bacteria is infected with bacteriophage at an m.o.i. of 0.2. Twenty-cell aliquots of the infected culture are withdrawn. How many phage-infected cells should be expected in each aliquot?

Solution 4.5 The number of phage-infected cells should equal the m.o.i. multiplied by the number of cells in the sample. The product represents an average number of infected cells per aliquot:

$$\frac{0.2 \text{ phage infections}}{\text{cell}} \times 20 \text{ cells} = 4 \text{ phage infections}$$

Therefore, in each 20-cell aliquot, there should be an average of four infected cells.

Problem 4.6 In a 20-cell aliquot from a culture infected at an m.o.i. of 0.2, what is the probability that no cells in that aliquot will be infected?

Solution 4.6 As shown in Problem 4.5, there should be, on average, four infected cells in a 20-cell aliquot. Using the Poisson distribution for the zero case, where r is equal to the number of successes (in this case, 0 infected cells is a success) and m is equal to the average number of infected cells per 20-cell aliquot (4), the Poisson distribution becomes

$$P_0 = \frac{e^{-4}4^0}{0!} = \frac{e^{-4}(1)}{(1)} = e^{-4}$$

To find e^{-4} on the calculator, enter **4**, **+/–**, then e^x. This gives a value of 0.018.

Therefore, the probability of finding no infected cells in a 20-cell aliquot taken from a culture infected at an m.o.i. of 0.2 is 0.018. This probability can also be expressed as a "1 in" number by taking its reciprocal:

$$\frac{1}{0.018} = 55.6$$

Or there is a 1-in-55.6 chance that a 20-cell aliquot will contain no infected cells.

Problem 4.7 In a 20-cell aliquot taken from a culture infected at an m.o.i. of 0.2, what is the probability of finding 12 infected cells?

Solution 4.7 The equation for the Poisson distribution can be used. In Problem 4.5, it was shown that the average number of infected cells in a 20-cell aliquot taken from such a culture is four and that this value represents the *m* factor in the Poisson distribution. The *r* factor for this problem, the number of successes, is 12. The Poisson distribution is then written

$$P_r = \frac{e^{-m}m^r}{r!}$$

$$P_{12} = \frac{e^{-4}4^{12}}{12!}$$

On the calculator, e^{-4} is found by entering **4**, **+/–**, then $\boldsymbol{e^x}$. This yields a value of 0.018. A value for 4^{12} is found on the calculator by entering **4**, $\boldsymbol{x^y}$, **1**, **2**, then **=**. This gives a value of 16,777,216. Twelve factorial (12!) is equal to the product of all integers from 1 to 12:

$$(12! = 12 \times 11 \times 10 \times 9 \times 8 \times 7 \times 6 \times 5 \times 4 \times 3 \times 2 \times 1)$$

On the calculator, this number is found by entering **1**, **2**, then ***x!***. This gives a value for 12! of 479,001,600. Placing these values into the equation for the Poisson distribution yields

$$P_{12} = \frac{e^{-4}4^{12}}{12!} = \frac{(0.018)(16{,}777{,}216)}{479{,}001{,}600} = \frac{301{,}990}{479{,}001{,}600} = 6.3 \times 10^{-4}$$

Therefore, the probability that 12 infected cells will be found in an aliquot of 20 cells taken from a culture infected at an m.o.i. of 0.2 is 0.00063. This value can be expressed as a "1 in" number by taking its reciprocal:

$$\frac{1}{6.3 \times 10^{-4}} = 1587.3$$

There is a 1-in-1587.3 chance that 12 infected cells will be found in such a 20-cell aliquot.

Problem 4.8 What is the probability that a sample of 20 cells taken from a culture infected at an m.o.i. of 0.2 will have four or more infected cells?

Solution 4.8 As a first step, the probabilities for the events not included (the probabilities for the cases in which zero, one, two, or three infected cells are found per 20-cell sample) are calculated. The probability of all possible infections (i.e., the probability of having no infected cells, of having one infected cell, two infected cells, three infected cells, four infected cells, five infected cells, etc.) should equal 1. If the probabilities of obtaining zero, one, two, and three infected cells are subtracted from 1.0, then the remainder will be equivalent to the probability of obtaining four or more infected cells. As shown in Problem 4.5, *m*, the average number of infected cells in a 20-cell aliquot from a culture infected at an m.o.i. of 0.2, is 4. The probability for the case in which no infected cells are present in a 20-cell aliquot was calculated in Problem 4.6 and was found to be 0.018.

The probability that the 20-cell aliquot will contain one infected cell is

$$P_1 = \frac{e^{-4}4^1}{1!} = \frac{(0.018)(4)}{1} = 0.072$$

The probability that the 20-cell aliquot will contain two infected cells is

$$P_2 = \frac{e^{-4}4^2}{2!} = \frac{(0.018)(16)}{2 \times 1} = \frac{0.288}{2} = 0.144$$

The probability that the 20-cell aliquot will contain three infected cells is

$$P_3 = \frac{e^{-4}4^3}{3!} = \frac{(0.018)(64)}{3 \times 2 \times 1} = \frac{1.152}{6} = 0.192$$

The probability of having four or more infected cells in a 20-cell aliquot is then

$$1 - P_0 - P_1 - P_2 - P_3 = 1 - 0.018 - 0.072 - 0.144 - 0.192 = 0.574$$

Therefore, 57.4% ($0.574 \times 100 = 57.4\%$) of the time, a 20-cell aliquot infected at an m.o.i. of 0.2 will contain four or more infected cells. Expressed as a "1 in" number, 1 in every 1.7 20-cell aliquots (the reciprocal of 0.574 = 1.7) will contain four or more infected cells.

Problem 4.9 What is the probability that in a 20-cell aliquot from a culture infected at an m.o.i. of 0.2, zero or one phage-infected cell will be found?

Solution 4.9 As shown in Problem 4.8, under these experimental conditions, the probability (P) of finding zero infected cells (P_0) is equal to 0.018 and the probability of finding one infected cell (P_1) is equal to 0.072. The probability of finding either zero or one phage-infected cell in a 20-cell aliquot is the sum of the two probabilities:

$$0.018 + 0.072 = 0.090$$

Therefore, 9% of the time (100 × 0.090 = 9%), either zero or one phage-infected cell will be found. Or, expressed as a "1 in" number, 1 in 11 20-cell aliquots will contain either zero or one infected cell (the reciprocal of 0.090 is equal to 11).

Problem 4.10 Following a period to allow for phage adsorption, cell culture samples are plated to determine the number of infected cells. It is found that 60% of the cells from the infected culture did not produce a burst of phage; i.e., they were not infected. What was the m.o.i. (the average number of phage-infected cells) of the culture?

Solution 4.10 For this problem, the equation for the Poisson distribution must be solved for m. Since the problem describes the zero case, the Poisson distribution,

$$P_r = \frac{e^{-m}m^r}{r!}$$

becomes

$$0.6 = \frac{e^{-m}m^0}{0!}$$

Since m^0 and 0! are both equal to 1, the equation is

$$0.6 = \frac{e^{-m}(1)}{1} = e^{-m}$$

To solve for m, the following relationship can be used: $P = e^{-m}$, which is equivalent to $m = -\ln P$. Solving for m yields

$$m = -\ln 0.6$$

To determine the natural log of 0.6 on a calculator, enter **.**, **6**, then **ln**. This gives a value, when rounded, of –0.51. Therefore,

$$m = -(-0.51)$$

Since the negative of a negative number is a positive value, $m = 0.51$.

Therefore, the m.o.i. of the culture was 0.51.

Measuring Phage Titer

The latter stage of bacteriophage development involves release of progeny phage particles from the cell. Different bacteriophages accomplish this step in different ways. Late during infection, the bacteriophage lambda encodes a protein, endolysin, which digests the bacterial cell wall. When the cell wall is critically weakened, the cell bursts open, releasing the 50 – 200 virus particles produced during the infection. The infected cell is killed by this process. Following replication of the M13 phage genome inside an infected cell, it is packaged into a protein coat as it passes through the cytoplasmic membrane on its way out of the cell. Mature M13 phage particles pass out of the cell individually rather than in a burst. Although M13 phage do not lyse infected cells, those harboring M13 grow more slowly than uninfected cells.

When plated with a lawn of susceptible cells, a bacteriophage will form a circular area of reduced turbidity called a **plaque**. Depending on the bacteriophage, a plaque may be clear (such as is formed by phage T1 or the clear mutants of lambda) or a plaque can be an area containing slow-growing cells (as is formed by M13 infection).

When a phage stock is diluted and plated with susceptible cells such that individual plaques can be clearly discerned on a bacterial lawn, it is possible to determine the concentration of virus particles within the stock. If the phage stock is diluted to a degree that individual and well-isolated plaques appear on the bacterial lawn, it is assumed that each plaque results from one phage infecting one cell. The process of determining phage concentration by dilution and plating with susceptible cells is called **titering** or the **plaque assay**. This method determines the number of viable phage particles in a stock suspension. A bacteriophage capable of productively infecting a cell is called a **plaque-forming unit** (**PFU**).

The following problems demonstrate how to calculate phage titer and to perform dilutions of phage stocks.

Problem 4.11 A bacteriophage stock is diluted in the following manner: 0.1 mL of the phage stock is diluted into 9.9 mL of dilution buffer (making a total volume in this first dilution tube of 10.0 mL). From this first dilution tube, 0.1 mL is withdrawn and diluted into a second tube containing 9.9 mL of dilution buffer. From the second dilution tube, 0.1 mL is taken and diluted into a third tube containing 9.9 mL of dilution buffer. From this third tube, 1.0 mL is withdrawn and added to 9.0 mL in a fourth tube. Finally, 0.1 mL is withdrawn from the fourth dilution tube and plated with 0.2 mL of susceptible cells in melted top agar. After incubating the plate overnight, 180 plaques are counted. What is the titer of the bacteriophage stock?

Solution 4.11 The series of dilutions described in this problem can be written so that each dilution step is represented as a fraction. To obtain the overall dilution factor, multiply all fractions.

$$\frac{0.1\ \text{mL}}{10\ \text{mL}} \times \frac{0.1\ \text{mL}}{10\ \text{mL}} \times \frac{0.1\ \text{mL}}{10\ \text{mL}} \times \frac{1.0\ \text{mL}}{10\ \text{mL}} \times 0.1\ \text{mL}$$

Expressing this series of fractions in scientific notation yields

$$(1\times10^{-2})\times(1\times10^{-2})\times(1\times10^{-2})\times(1\times10^{-1})\times(1\times10^{-1}\ \text{mL}) = 1\times10^{-8}\ \text{mL}$$

To obtain the concentration of phage in the stock suspension, divide the number of plaques counted by the dilution factor.

$$\frac{180\ \text{PFU}}{1\times10^{-8}\ \text{mL}} = 1.8\times10^{10}\ \text{PFU/mL}$$

Therefore, the bacteriophage stock has a titer of 1.8×10^{10} PFU/mL

Diluting Bacteriophage

It is often necessary to dilute a bacteriophage stock so that the proper amount (and a convenient volume) of virus can be added to a culture. The following considerations should be taken into account when planning a dilution scheme.

- A 0.1-mL aliquot taken from the last dilution tube is a convenient volume to plate with susceptible cells. This would comprise 1×10^{-1} of the 2×10^{-7} dilution factor.
- A 0.1-mL aliquot is a convenient volume to remove from the phage stock into the first dilution tube. If the first dilution tube contains 9.9 mL of buffer, then this would account for 1×10^{-2} of the 2×10^{-7} dilution factor.
- Since the significant digit of the dilution factor is 2 (the 2 in 2×10^{-7}), this number must be brought into the dilution series. This can be accomplished by making a 0.2 mL/10 mL dilution (equivalent to a dilution factor of 2×10^{-2}). Therefore, 0.2 mL taken from the first dilution tube can be added to a second dilution tube, containing 9.8 mL of buffer.

Problem 4.12 A phage stock has a concentration of 2.5×10^{9} PFU/mL. How can the stock be diluted and plated to give 500 plaques on a plate?

Solution 4.12 This problem can be tackled by taking the reverse approach of that taken for Problem 4.11. If x represents the dilution factor required to form 500 plaques per plate from a phage stock having a concentration of 2.5×10^{9} PFU/mL, then the following equation can be written:

$$\frac{500\ \text{PFU}}{x\ \text{mL}} = \frac{2.5\times10^{9}\ \text{PFU}}{\text{mL}}$$

Solving for x gives

$$500 \text{ PFU} = \frac{(2.5 \times 10^9 \text{ PFU})(x \text{ mL})}{\text{mL}}$$

$$\frac{500 \text{ PFU}}{2.5 \times 10^9 \text{ PFU}} = x = 2 \times 10^{-7}$$

Therefore, the phage stock must be diluted by 2×10^{-7}.

The phage can be diluted as follows:

$$(1 \times 10^{-2}) \times (2 \times 10^{-2}) \times (1 \times 10^{-1} \text{ mL}) = 2 \times 10^{-5} \text{ mL}$$

The remaining amount of the dilution factor to be accounted for is

$$(2 \times 10^{-5}) \times (x) = 2 \times 10^{-7}$$

Solving for x yields

$$x = \frac{2 \times 10^{-7}}{2 \times 10^{-5}} = 1 \times 10^{-2}$$

This remaining part of the dilution series can be accomplished by taking 0.1 mL into 9.9 mL of buffer and placing this dilution somewhere in the series.

Therefore, if a phage stock at a concentration of 2.5×10^9 PFU/mL is diluted as follows:

$$\frac{0.1 \text{ mL}}{10 \text{ mL}} \times \frac{0.2 \text{ mL}}{10 \text{ mL}} \times \frac{0.1 \text{ mL}}{10 \text{ mL}} \times 0.1 \text{ mL plated}$$

then 500 plaques should appear on the plated lawn.

Measuring Burst Size

Mutation of either the phage or the bacterial genome or changes in the conditions under which infection occurs can alter the phage/host interaction and the efficiency with which phage replication occurs. Measuring the number of progeny bacteriophage particles produced within and released from an infected cell is a simple way to gauge overall gene expression of the infecting phage. The number of mature virus particles released from an infected cell is called the **burst size.**

To perform an experiment measuring burst size, for example, phage are added to susceptible cells at an m.o.i. of 0.1 and allowed to adsorb for 10 – 30 minutes on ice. The cell/phage suspension is diluted into a large volume with cold growth media and centrifuged to pellet the cells. The supernatant is poured off to remove unadsorbed phage. The pelleted cells are resuspended in cold media and then diluted into a growth flask containing prewarmed media. A sample is taken immediately and assayed for plaque-forming units (PFUs). The plaques arising from this titration represent the number of infected cells [termed **infective centers (ICs)**]; each infected cell gives rise to one plaque. After a period of time sufficient for cell lysis to occur, phage are titered again for PFUs. The burst size is calculated as plaque-forming units released per infective center (PFUs/IC).

Problem 4.13 In a burst size experiment, phage are added to 2×10^8 cells at an m.o.i. of 0.1 and allowed to adsorb to susceptible cells for 20 minutes. Dilution and centrifugation is performed to remove unadsorbed phage. The pelleted infected cells are resuspended in 10 mL of tryptone broth. At this point, infective centers are assayed by diluting an aliquot of the infected cells as follows:

$$\frac{0.1 \text{ mL}}{10 \text{ mL}} \times \frac{0.1 \text{ mL}}{10 \text{ mL}} \times 0.1 \text{ mL plated}$$

The resuspended infected cells are then shaken at 37°C for 90 minutes. The number of PFUs are then assayed by diluting an aliquot as follows:

$$\frac{0.1 \text{ mL}}{10 \text{ mL}} \times \frac{0.1 \text{ mL}}{10 \text{ mL}} \times \frac{1 \text{ mL}}{10 \text{ mL}} \times \frac{0.5 \text{ mL}}{10 \text{ mL}} \times 0.1 \text{ mL plated}$$

From the assay of infective centers, 200 plaques are counted on the bacterial lawn. From the assay of PFUs following the 90-minute incubation, 150 plaques are counted from the diluted sample. What is the burst size, expressed as PFUs/IC?

Solution 4.13 The number of infective centers is determined by dividing the number of plaques obtained following dilution by the dilution factor. The dilution scheme to obtain infective centers was

$$\frac{0.1 \text{ mL}}{10 \text{ mL}} \times \frac{0.1 \text{ mL}}{10 \text{ mL}} \times 0.1 \text{ mL plated}$$

This is equivalent to the expression

$$(1 \times 10^{-2})(1 \times 10^{-2})(1 \times 10^{-1} \text{ mL}) = 1 \times 10^{-5} \text{ mL}$$

The number of infective centers is then equal to

$$\frac{200 \text{ PFUs}}{1\times10^{-5} \text{ mL}} = 2\times10^{7} \text{ PFUs/mL } = 2\times10^{7} \text{ ICs/mL}$$

The number of PFUs following the 90-minute incubation was obtained by diluting the sample in the following manner:

$$\frac{0.1 \text{ mL}}{10 \text{ mL}} \times \frac{0.1 \text{ mL}}{10 \text{ mL}} \times \frac{1 \text{ mL}}{10 \text{ mL}} \times \frac{0.5 \text{ mL}}{10 \text{ mL}} \times 0.1 \text{ mL plated}$$

This is equivalent to 5×10^{-8} as shown below.

$$(1\times10^{-2})(1\times10^{-2})(1\times10^{-1})(5\times10^{-2})(1\times10^{-1} \text{ mL}) = 5\times10^{-8} \text{ mL}$$

Since 150 plaques were obtained from this dilution, the concentration of phage in the culture after 90 minutes is

$$\frac{150 \text{ PFUs}}{5\times10^{-8} \text{ mL}} = 3\times10^{9} \text{ PFUs/mL}$$

The burst size is then calculated as the concentration of phage following 90-minute incubation divided by the concentration of infective centers:

$$\frac{3\times10^{9} \text{ PFUs/mL}}{2\times10^{7} \text{ ICs/mL}} = \frac{3\times10^{9} \text{ PFUs}}{\text{mL}} \times \frac{\text{mL}}{2\times10^{7} \text{ ICs}}$$

$$= \frac{3\times10^{9} \text{ PFUs}}{2\times10^{7} \text{ ICs}} = 150 \text{ PFUs/IC}$$

Note: *To perform this calculation, a fraction (3×10^{9} PFUs/mL) is divided by a fraction (2×10^{7} ICs/mL). The relationship described in the box can be used in such a situation.*

Dividing a fraction by a fraction is the same as multiplying the numerator fraction by the reciprocal of the denominator fraction. A phrase frequently used to describe this action is to "invert and multiply." Therefore,

$$\frac{\frac{1}{a}}{\frac{1}{b}} = \frac{1}{a} \times \frac{b}{1} = \frac{b}{a} \quad \text{and} \quad \frac{\frac{a}{1}}{\frac{b}{1}} = \frac{a}{1} \times \frac{1}{b} = \frac{a}{b}$$

Quantitation of Nucleic Acids

5

Quantitation of Nucleic Acids by Ultraviolet Spectroscopy

Any experiment requiring manipulation of a nucleic acid most likely also requires its accurate quantitation to ensure optimal and reproducible results. The nitrogenous bases positioned along a nucleic acid strand absorb ultraviolet (UV) light at a wavelength of 260 nm (nanometers); at this wavelength, light absorption is proportional to nucleic acid concentration. This relationship is so well characterized that UV absorption is used to accurately determine the concentration of nucleic acids in solution. The relationship between DNA concentration and absorption is linear up to an absorption at 260 nm (A_{260}) of 2 (Figure 5.1). For measuring the absorption of a nucleic acid solution in a spectrophotometer, most molecular biology laboratories will use quartz cuvettes with a width through which the light beam will travel 1 cm. Therefore, all discussions in this section assume a 1-cm light path.

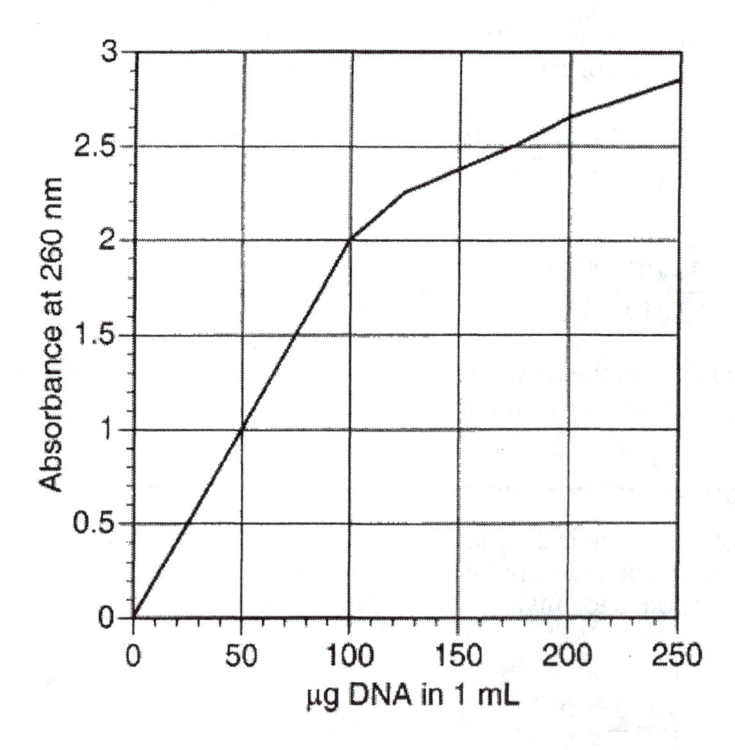

Figure 5.1 The concentration of DNA and absorbance at 260 nm is a linear relationship up to an A_{260} value of approximately 2.

Determining the Concentration of Double-Stranded DNA

Applications requiring quantitation of double-stranded DNAs include protocols utilizing plasmids, viruses, or genomes. Quantitation is typically performed by taking absorbance measurements at 260 nm, 280 nm, and 320 nm. Absorbance at 260 nm is used to specifically detect the nucleic acid component of a solution. Absorbance at 280 nm is used to detect the presence of protein (since tryptophan residues absorb at this wavelength). Absorbance at 320 nm is used to detect any insoluble light-scattering components. A spectrophotometer capable of providing a scan from 200 to 320 nm will yield maximum relevant information (Figure 5.2).

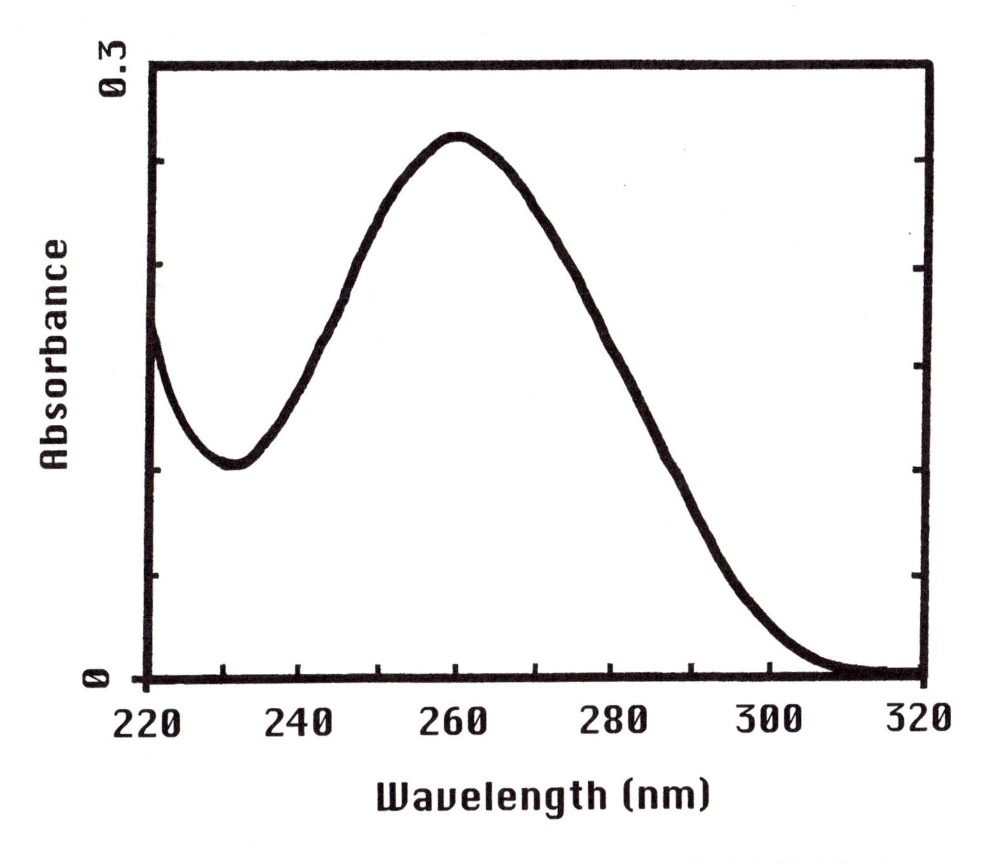

Figure 5.2 A typical spectrophotometric scan of double-stranded DNA. Maximum absorbance occurs at 260 nm. Absorbance at 230 nm is an indication of salt in the sample. If the sample is optically clear, it should give a very low reading at 320 nm, which is in the visible wavelength region of the spectrum.

For nucleic acids purified from a biological source (as opposed to those made synthetically), calculating the ratio of the readings obtained at 260 nm and 280 nm can give an estimate of protein contamination. Pure DNA free of protein contamination will have an A_{260}/A_{280} ratio close to 1.8. If phenol or protein contamination is present in the DNA prep, the A_{260}/A_{280} ratio will be less than 1.8. If RNA is present in the DNA prep, the A_{260}/A_{280} ratio may be greater than 1.8. Pure RNA preparations will have an A_{260}/A_{280} ratio close to 2.0.

At 260 nm, DNA concentrations as low as 2 μg/mL can be detected. A solution of DNA with a concentration of 50 μg/mL will have an absorbance at 260 nm equal to 1.0. Written as an equation, this relationship is

$$1\ A_{260} \text{ unit of double-stranded DNA} = 50\ \mu\text{g DNA/mL}$$

Absorbance and **optical density (OD)** are terms often used interchangeably. The foregoing relationship can also be written as a conversion factor:

$$\frac{50\ \mu\text{g DNA/mL}}{1.0\ \text{OD}}$$

The following problems use this relationship.

Problem 5.1 From a small culture, you have purified the DNA of a recombinant plasmid. You have resuspended the DNA in a volume of 50 μL. You dilute 20 μL of the purified DNA sample into a total volume of 1000 μL distilled water. You measure the absorbance of this diluted sample at 260 nm and 280 nm and obtain the following readings.

$$A_{260} = 0.550$$

$$A_{280} = 0.324$$

a) What is the DNA concentration of the 50-μL plasmid prep?

b) How much total DNA was purified by the plasmid prep procedure?

c) What is the A_{260}/A_{280} ratio of the purified DNA?

Solution 5.1(a) This problem can be solved by setting up the following ratio.

$$\frac{x\ \mu\text{g DNA/mL}}{0.550\ \text{OD}} = \frac{50\ \mu\text{g DNA/mL}}{1.0\ \text{OD}}$$

$$x\ \mu\text{g DNA/mL} = \frac{(0.550\ \text{OD})(50\ \mu\text{g DNA/mL})}{1.0\ \text{OD}}$$

$$x = 27.5\ \mu\text{g DNA/mL}$$

This amount represents the concentration of the diluted DNA solution that was used for the spectrophotometer. To determine the concentration of DNA in the original 50-μL plasmid prep, this value must be divided by the dilution factor. A 20-μL sample of the plasmid prep DNA was diluted into water for a total diluted volume of 1000 μL.

$$\frac{27.5\ \mu\text{g DNA/mL}}{\frac{20}{1000}} = 27.5\ \mu\text{g DNA/mL} \times \frac{1000}{20}$$

$$= \frac{27{,}500\ \mu\text{g DNA/mL}}{20} = 1375\ \mu\text{g DNA/mL}$$

Therefore, the original 50-μL plasmid prep has a concentration of 1375 μg DNA/mL. To bring this value to an amount of DNA per μL, it can be multiplied by the conversion factor, 1 mL/1000 μL:

$$\frac{1375\ \mu\text{g DNA}}{\text{mL}} \times \frac{1\ \text{mL}}{1000\ \mu\text{L}} = 1.35\ \mu\text{g DNA/}\mu\text{L}$$

Therefore, the 50-μL plasmid prep has a concentration of 1.35 μg DNA/μL.

Solution 5.1(b) A total amount of DNA recovered by the plasmid prep procedure can be calculated by multiplying the DNA concentration obtained earlier by the volume containing the recovered DNA:

$$\frac{1375\ \mu\text{g DNA}}{\text{mL}} \times \frac{1\ \text{mL}}{1000\ \mu\text{L}} \times 50\ \mu\text{L} = \frac{(1375)(50)}{1000} = 68.75\ \mu\text{g DNA}$$

Therefore, the original 50-μL plasmid prep contained a total of 68.75 μg DNA. However, since 20 μL was used for diluting and reading in the spectrophotometer, only 30 μL of sample prep remains. Therefore, the total remaining amount of DNA is calculated by multiplying the remaining volume by the concentration:

$$30\ \mu\text{L} \times \frac{1.35\ \mu\text{g DNA}}{\mu\text{L}} = 40.5\ \mu\text{g DNA}$$

Therefore, although 68.75 μg of DNA were recovered from the plasmid prep procedure, you used up some of it for spectrophotometry and you now have 40.5 μg DNA remaining.

Solution 5.1(c) The A_{260}/A_{280} ratio is

$$\frac{0.550}{0.323} = 1.703$$

Using Absorbance and an Extinction Coefficient to Calculate Double-Stranded DNA Concentration

At a neutral pH and assuming a G+C DNA content of 50%, a DNA solution having a concentration of 1 mg/mL will have an absorption in a 1-cm light path at 260 nm (A_{260}) of 20. For most applications in molecular biology, G+C content, unless very heavily skewed, need not be a consideration when quantitating high-molecular-weight double-stranded DNA. The absorption value of 20 for a 1-mg DNA/mL solution is referred to as DNA's **extinction coefficient**. It is represented by the symbol E or ε. The term **extinction coefficient** is used interchangeably with the **absorption constant** or **absorption coefficient**. The formula that describes the relationship between absorption at 260 nm (A_{260}), concentration (c, in mg/mL), the light path length (l) of the cuvette (in centimeters), and the extinction coefficient at 260 nm (E_{260}) for a 1-cm light path is

$$A_{260} = E_{260} lc$$

This relationship is known as **Beer's law**. Since the light path, l, is 1, this equation becomes

$$A_{260} = E_{260} c$$

Rearranging the equation, the concentration of the nucleic acid, c, becomes

$$c = \frac{A_{260}}{E_{260}}$$

Problem 5.2 A DNA solution has an A_{260} value of 0.5. What is the DNA concentration in micrograms DNA/mL?

Solution 5.2 The answer can be obtained using the equation for Beer's Law.

$$c = \frac{A_{260}}{E_{260}}$$

$$c = \frac{0.5}{20} = 0.025 \text{ mg DNA/mL}$$

$$0.025 \text{ mg DNA/mL} \times \frac{1000\ \mu\text{g}}{\text{mg}} = 25\ \mu\text{g DNA/mL}$$

Therefore, the sample has a concentration of 25 μg DNA/mL.

Problem 5.3 A DNA solution has an A_{260} value of 1.0. What is the DNA concentration in micrograms DNA/mL?

Solution 5.3 The answer can be obtained using the equation for Beer's Law.

$$c = \frac{A_{260}}{E_{260}}$$

$$c = \frac{1.0}{20} = 0.05 \text{ mg DNA/mL}$$

$$0.05 \text{ mg DNA/mL} \times \frac{1000\ \mu\text{g}}{\text{mg}} = 50\ \mu\text{g DNA/mL}$$

Therefore, the concentration of the solution is 50 μg DNA/mL. Notice that this value is the one described earlier.

$$1\ A_{260} \text{ unit of double-stranded DNA} = 50\ \mu\text{g DNA/mL}$$

Calculating DNA Concentration as a Millimolar (m*M*) Amount

The extinction coefficient (E_{260}) for a 1 m*M* solution of double-stranded DNA is 6.7. This value can be used to calculate the molarity of a solution of DNA.

Problem 5.4 A solution of DNA has an absorbance at 260 nm of 0.212. What is the concentration of the DNA solution expressed as millimolarity?

Solution 5.4 This problem can be solved by setting up a relationship of ratios such that it is read "1 m*M* is to 6.7 OD as *x* m*M* is to 0.212 OD."

$$\frac{1\ \text{m}M}{6.7\ \text{OD}} = \frac{x\ \text{m}M}{0.212\ \text{OD}}$$

$$\frac{(1\ \text{m}M)(0.212\ \text{OD})}{6.7\ \text{OD}} = x\ \text{m}M$$

$$0.03\ \text{m}M = x$$

Therefore, a DNA solution with an A_{260} of 0.212 has a concentration of 0.03 m*M*.

Problem 5.5 A solution of DNA has an absorbance at 260 nm of 1.00. What is its concentration expressed as millimolarity?

Solution 5.5 This problem can be solved using ratios, with one of those ratios being the relationship of a 1 m*M* solution of double-stranded DNA to the extinction coefficient 6.7.

$$\frac{x\ \text{m}M}{1.00\ \text{OD}} = \frac{1\ \text{m}M}{6.7\ \text{OD}}$$

$$x\ \text{m}M = \frac{(1\ \text{m}M)(1.00\ \text{OD})}{6.7\ \text{OD}} = 0.15\ \text{m}M$$

Therefore, a solution of double-stranded DNA with an A_{260} of 1.00 has a concentration of 0.15 m*M*. This relationship,

$$1.0\ A_{260}\ \text{of double-stranded DNA} = 0.15\ \text{m}M$$

has frequent use in the laboratory.

Problem 5.6 A solution of DNA has a concentration of 0.03 m*M*. What is its concentration expressed as pmol/µL?

Solution 5.6 A 0.03 m*M* solution, by definition, has a concentration of 0.03 millimoles per liter. A series of conversion factors is used to cancel terms and to transform a concentration expressed as millimolarity to one expressed as pmol/µL:

$$\frac{0.03 \text{ mmol}}{\text{L}} \times \frac{1 \text{ Liter}}{1 \times 10^6 \ \mu\text{L}} \times \frac{1 \times 10^9 \text{ pmol}}{\text{mmol}} = 30 \text{ pmol/mL}$$

Therefore, a 0.03 m*M* DNA solution has a concentration of 30 pmol DNA/µL.

Determining the Concentration of Single-Stranded DNA Molecules

Single-Stranded DNA Concentration Expressed in µg/mL

To determine the concentration of single-stranded DNA (ssDNA) as a µg/mL amount, the following conversion factor is used:

$$1 \text{ OD of ssDNA} = 33 \ \mu\text{g/mL}$$

Problem 5.7 Single-stranded DNA isolated from M13mp18, a derivative of bacteriophage M13 used in cloning and DNA sequencing applications, is diluted 10 µL into a total volume of 1000 µL water. The absorbance of this diluted sample is read at 260 nm and an A_{260} value of 0.325 is obtained. What is its concentration in µg/mL?

Solution 5.7 This problem can be solved by setting up a ratio, with the variable x representing the concentration in µg/mL for the diluted sample. The equation can be read "x µg/mL is to 0.125 OD as 33 µg/mL is to 1 OD." Once x is obtained, the concentration of the stock DNA can be determined by multiplying the concentration of the diluted sample by the dilution factor.

$$\frac{x \ \mu\text{g/mL}}{0.125 \text{ OD}} = \frac{33 \ \mu\text{g/mL}}{1 \text{ OD}}$$

$$x \ \mu\text{g/mL} = 0.125 \times 33 \ \mu\text{g/mL} = 4.125 \ \mu\text{g/mL}$$

Therefore, the concentration of the diluted sample is 4.125 µg/mL. To determine the concentration of the M13mp18 DNA stock solution, this value must be multiplied by the dilution factor:

$$4.125 \ \mu g/mL \times \frac{1000 \ \mu L}{10 \ \mu L} = \frac{4125 \ \mu g/mL}{10} = 412.5 \ \mu g/mL$$

Therefore, the stock of M13mp18 DNA has a concentration of 412.5 μg/mL.

Determining the Concentration of High-Molecular-Weight Single-Stranded DNA in pmol/μL

The concentration of high-molecular-weight ssDNA can be expressed as a pmol/μL amount by first determining how many micrograms of the ssDNA are equivalent to 1 picomole. To do this, we use the average molecular weight of a deoxynucleotide in a DNA strand. For single-stranded DNA, it is taken to be 330 daltons. This value is then used as a conversion factor to bring the concentration of ssDNA expressed as μg/mL to a concentration expressed in pmol/μL.

> The unit **dalton** is defined as 1/12 the mass of the carbon-12 atom. It is used interchangeably with "molecular weight," a quantity expressed as grams/mole. That is, there are 330 grams of nucleotide per mole of nucleotide.

Problem 5.8 A stock of M13mp18 DNA has a concentration of 412.5 μg/mL. What is this concentration expressed in pmol/μL?

Solution 5.8 The cloning vector M13mp18 is 7250 nucleotides (nts) in length. To express this as μg/pmol, the following relationship is set up, in which a series of conversion factors is used to cancel terms:

$$7250 \text{ nts} \times \frac{330 \text{ g/mol}}{\text{nt}} \times \frac{1 \times 10^6 \ \mu g}{g} \times \frac{\text{mole}}{1 \times 10^{12} \text{ pmol}} = 2.39 \ \mu g/\text{pmol}$$

Therefore, 2.39 μg of a 7250-nucleotide-long ssDNA molecule are equivalent to 1 pmol. This value can now be used to convert μg/mL to pmol/μL:

$$\frac{412.5 \ \mu g}{mL} \times \frac{1 \text{ pmol}}{2.39 \ \mu g} \times \frac{1 \ mL}{1000 \ \mu L} = 0.17 \ \text{pmol}/\mu L$$

Therefore, the M13mp18 DNA stock has a concentration of 0.17 pmol/μL.

Expressing ssDNA Concentration as a Millimolarity Amount

The extinction coefficient (E_{260}) for a 1 mM solution of single-stranded DNA is 8.5. This value can be used to determine the millimolarity concentration of any single-stranded DNA solution from its absorbance.

Problem 5.9 A 1-mL sample of single-stranded DNA has an absorbance of 0.285. What is its m*M* concentration?

Solution 5.9 The following relationship can be used to determine the millimolarity concentration.

$$\frac{x \text{ m}M}{0.285 \text{ OD}} = \frac{1 \text{ m}M}{8.5 \text{ OD}}$$

$$x \text{ m}M = \frac{(1 \text{ m}M)(0.285 \text{ OD})}{8.5 \text{ OD}} = 0.03 \text{ m}M$$

Therefore, a solution of single-stranded DNA with an absorbance of 0.285 has a concentration of 0.03 m*M*.

Oligonucleotide Quantitation

Optical Density (OD) Units

Many laboratories express an amount of an oligonucleotide in terms of **optical density (OD) units**. An **OD unit** is the amount of oligonucleotide dissolved in 1.0 mL giving an A_{260} of 1.00 in a cuvette with a 1-cm light path length. It is calculated by the equation

$$\text{OD units} = (A_{260}) \times (\text{oligonucleotide volume}) \times (\text{dilution factor})$$

Problem 5.10 Following its synthesis, an oligonucleotide is dissolved in 1.5 mL of water. You dilute 50 µL of the oligonucleotide into a total volume of 1000 µL and read the absorbance of the diluted sample at 260 nm. An A_{260} of 0.264 is obtained. How many OD units are present in the 1.5 mL of oligonucleotide stock?

Solution 5.10 Using the formula just given, the number of OD units is

$$\text{OD units} = 0.264 \times 1.5 \text{ mL} \times \frac{1000 \text{ µL}}{50 \text{ µL}}$$

$$= \frac{396}{50} = 7.92 \text{ OD units}$$

Therefore, the 1.5 mL solution contains 7.92 OD units of oligonucleotide.

Expressing an Oligonucleotide's Concentration in μg/mL

An A_{260} reading can be converted into a concentration expressed as μg/mL using the extinction coefficient for single-stranded DNA (ssDNA) of 1 mL/33 μg for a 1-cm light path. In other words, a solution of ssDNA with an A_{260} value of 1.0 (1.0 OD unit) contains 33 μg of ssDNA per milliliter. Written as equation, this relationship is

$$1 \text{ OD unit} = 33 \text{ μg ssDNA/mL}$$

Problem 5.11 In the previous problem, a diluted oligonucleotide gave an A_{260} reading of 0.264. What is the concentration of the oligonucleotide in μg DNA/mL?

Solution 5.11 This problem can be solved using the following ratio.

$$\frac{x \text{ μg ssDNA/mL}}{0.264 \text{ OD}} = \frac{33 \text{ μg ssDNA/mL}}{1 \text{ OD}}$$

$$x \text{ μg ssDNA} = \frac{(0.264 \text{ OD})(33 \text{ μg ssDNA/mL})}{1 \text{ OD}}$$

$$x = 8.712 \text{ μg ssDNA/mL}$$

This value represents the concentration of oligonucleotide in the diluted sample used for spectrophotometry. The concentration of the oligonucleotide stock solution is obtained by multiplying this value by the dilution factor (the inverse of the dilution):

$$8.712 \text{ μg ssDNA/mL} \times \frac{1000 \text{ μL}}{50 \text{ μL}} = 174.24 \text{ μg ssDNA/mL}$$

Therefore, the concentration of the oligonucleotide stock solution is 174.24 μg ssDNA/mL.

Oligonucleotide Concentration Expressed in pmol/μL

Many applications in molecular biology require that a certain number of picomoles (pmol) of oligonucleotide be added to a reaction. This is true in the case of fluorescent DNA sequencing, for example, which, for some protocols, requires 3.2 picomoles of oligonucleotide primer per sequencing reaction. An oligonucleotide's concentration can be calculated from an A_{260} value by using the formula

$$C = \frac{A_{260} \times 100}{(1.54 \times n\text{A}) + (0.75 \times n\text{C}) + (1.17 \times n\text{G}) + (0.92 \times n\text{T})} \times \text{dilution factor}$$

In this formula, the concentration, C, is calculated as picomoles per microliter (pmol/μL). The denominator consists of the sum of the extinction coefficents of each base multiplied by the number of times that base appears in the oligonucleotide.

Extinction Coefficients for the Bases of DNA

The extinction coefficients for the bases are usually expressed in liters per mole or as the absorbance for a 1 *M* solution at the wavelength where the base exhibits maximum absorbance. Either way, the absolute value is the same. The extinction coefficient for dATP, for example, is 15,400 L/mol, or a 1 *M* solution (at pH 7) of dATP will have an absorbance of 15,400 at the wavelength of its maximum absorbance. The amount 15,400 L/mol is equivalent to 0.0154 μL/pmol, as shown by the following conversion:

$$\frac{15{,}400 \text{ L}}{\text{mole}} \times \frac{1 \times 10^{6} \ \mu\text{L}}{1 \text{ L}} \times \frac{1 \text{ mole}}{1 \times 10^{12} \text{ pmol}} = \frac{0.0154 \ \mu\text{L}}{\text{pmol}}$$

The earlier formula to calculate oligonucleotide concentration uses the value 1.54 as the extinction coefficient for dATP. This value is obtained by multiplying the expression by 100/100 (which is the same as multiplying by 1). This manipulation leaves a “100” in the numerator and makes the equation more manageable.

Problem 5.12 An oligonucleotide with the sequence GAACTACGTTCGATCAAT is suspended in 750 μL of water. A 20-μL aliquot is diluted to a final volume of 1000 μL with water, and the absorbance at 260 nm is determined for the diluted sample. An optical density (OD) of 0.242 is obtained. What is the concentration of the oligonucleotide stock solution in pmol/μL?

Solution 5.12 The oligoncleotide contains six A residues, four C residues, three G residues, and five T residues. Placing these values into the equation given earlier yields the following result.

$$C = \frac{A_{260} \times 100}{(1.54 \times n\text{A}) + (0.75 \times n\text{C}) + (1.17 \times n\text{G}) + (0.92 \times n\text{T})} \times \text{dilution factor}$$

$$\text{pmol}/\mu\text{L} = \frac{0.242 \times 100}{(1.54 \times 6) + (0.75 \times 4) + (1.17 \times 3) + (0.92 \times 5)} \times \frac{1000\ \mu\text{L}}{20\ \mu\text{L}}$$

$$\text{pmol}/\mu\text{L} = \frac{24.2}{9.24 + 3.00 + 3.51 + 4.60} \times \frac{1000}{20}$$

$$\text{pmol}/\mu\text{L} = \frac{24.2}{20.35} \times \frac{1000}{20} = \frac{24{,}200}{407} = 59.46$$

Therefore, the oligonucleotide stock solution has a concentration of 59.46 pmol/μL.

Problem 5.13 An oligonucleotide stock has a concentration of 60 pmol/μL. You wish to use 3.2 picomoles of the oligonucleotide as a primer in a DNA sequencing reaction. How many microliters of oligo stock will give you the desired 3.2 picomoles?

Solution 5.13 The answer can be obtained in the following manner:

$$\frac{60\ \text{pmol}}{\mu\text{L}} \times x\ \mu\text{L} = 3.2\ \text{pmol}$$

Setting up the equation in this way allows cancellation of units to give us the number of picomoles. Solving then for x yields

$$x = \frac{3.2\ \text{pmol}}{60\ \text{pmol}/\mu\text{L}} = 0.05\ \mu\text{L}$$

Therefore, to deliver 3.2 picomoles of oligonucleotide primer to a reaction, you need to take a 0.05-μL aliquot of the 60 pmol/μL oligonucleotide stock. However, delivering that small of a volume with a standard laboratory micropipet is neither practical nor accurate. It is best to dilute the sample to a concentration such that the desired amount of DNA is contained within a volume that can be accurately delivered by your pipettor system. For example, say that you wanted to deliver 3.2 picomoles of oligonucleotide in a 2-μL volume. You can then ask, "If I take a certain specified and convenient volume from the oligonucleotide stock solution, say 5 μL, into what volume of water (or TE dilution buffer) do I need to dilute that 5 μL to give me a concentration such that 3.2 picomoles of oligonucleotide are contained in a 2-μL aliquot?" That problem can be set up in the following way.

$$\frac{60\text{ pmol}}{\mu\text{L}} \times \frac{5\ \mu\text{L}}{x\ \mu\text{L}} = \frac{3.2\text{ pmol}}{2\ \mu\text{L}}$$

$$\frac{300\text{ pmol}}{x\ \mu\text{L}} = \frac{3.2\text{ pmol}}{2\ \mu\text{L}}$$

$$300\text{ pmol} = \frac{3.2x\text{ pmol}}{2}$$

$$600\text{ pmol} = 3.2x\text{ pmol}$$

$$\frac{600\text{ pmol}}{3.2\text{ pmol}} = x$$

$$187.5 = x$$

Therefore, if 5 μL is taken from the 60 pmol/μL oligonucleotide stock solution and diluted to a final volume of 187.5 μL, then 2 μL of this diluted sample will contain 3.2 picomoles of oligonucleotide.

Measuring RNA Concentration

The following relationship is used for quantitating RNA:

$$1\ A_{260}\text{ unit of RNA} = 40\ \mu\text{g/mL}$$

Problem 5.14 Forty microliters of a stock solution of RNA is diluted with water to give a final volume of 1000 μL. The diluted sample has an absorbance at 260 nm of 0.142. What is the concentration of the RNA stock solution in μg/mL?

Solution 5.14 Ratios are set up that read "x μg RNA/mL is to 0.142 OD as 40 μg RNA/mL is to 1.0 OD."

$$\frac{x\ \mu\text{g RNA/mL}}{0.142\ \text{OD}} = \frac{40\ \mu\text{g RNA/mL}}{1\ \text{OD}}$$

$$x\ \mu\text{g RNA/mL} = \frac{(0.142\ \text{OD})(40\ \mu\text{g RNA/mL})}{1\ \text{OD}} = 5.7\ \mu\text{g RNA/mL}$$

This value represents the concentration of the diluted sample. To obtain the concentration of the RNA stock solution, it must be multiplied by the dilution factor:

$$5.7\ \mu\text{g RNA/mL} \times \frac{1000\ \mu\text{L}}{40\ \mu\text{L}} = 142.5\ \mu\text{g RNA/mL}$$

Therefore, the RNA stock solution has a concentration of 142.5 μg RNA/mL.

Molecular Weight, Molarity, and Nucleic Acid Length

The average molecular weight of a DNA base is approximately 330 daltons (or 330 grams/mole). The average molecular weight of a DNA base pair is twice this, approximately 660 daltons (or 660 grams/mole). These values can be used to calculate how much DNA is present in any biological source.

For small single-stranded DNA molecules, such as synthetic oligonucleotides, the molecular weight of the individual nucleotides can be added to determine the strand's total molecular weight (MW) according to the following formula:

$$\text{MW} = (n_{\text{A}} \times 335.2) + (n_{\text{C}} \times 311.2) + (n_{\text{G}} \times 351.2) + (n_{\text{T}} \times 326.2) + P$$

where n_X is the number of nucleotides of A, C, G, or T in the oligonucleotide and P is equal to −101.0 for dephosphorylated (lacking an end phosphate group) or 40.0 for phosphorylated oligonucleotides.

The following problems demonstrate how molecular weight, molarity, and nucleic acid length relate to DNA quantity.

Problem 5.15 How much genomic DNA is present inside a single human diploid nucleated cell? Express the answer in picograms.

Solution 5.15 The first step in solving this problem is to calculate how many grams a single base pair weighs. This weight will then be multiplied by the number of base pairs in a single diploid cell. The first calculation uses Avogadro's number:

$$\frac{660\ \text{g}}{\text{mol}} \times \frac{1\ \text{mol}}{6.023 \times 10^{23}\ \text{bp}} = 1.1 \times 10^{-21}\ \text{g/bp}$$

Therefore, a single base pair weighs 1.1×10^{-21} grams.

There are approximately 6×10^{9} base pairs per diploid human nucleated cell. This can be converted to a picogram amount by using the conversion factor 1×10^{12} picograms/gram, as shown in the following equation:

$$6 \times 10^{9}\ \text{bp} \times \frac{1.1 \times 10^{-21}\ \text{g}}{\text{bp}} \times \frac{1 \times 10^{12}\ \text{pg}}{\text{g}} = 7\ \text{pg}$$

Therefore, a single diploid human cell contains 7 picograms of genomic DNA.

Problem 5.16 You have 5 mL of bacteriophage lambda stock having a concentration of 2.5×10^{11} phage/mL. You will purify lambda DNA using the entire stock. Assuming 80% recovery, how many micrograms of lambda DNA will you obtain?

Solution 5.16 First, calculate the total number of phage particles by multiplying the stock's concentration by its total volume:

$$\frac{2.5 \times 10^{11}\ \text{phage}}{\text{mL}} \times 5\ \text{mL} = 1.25 \times 10^{12}\ \text{phage}$$

In the previous problem, we found that one base pair weighs 1.1×10^{-21} grams. The bacteriophage lambda is 48,502 bp in length. With these values, the quantity of DNA contained in the 5 mL of lambda phage stock can be calculated:

$$1.25 \times 10^{12}\ \text{phage} \times \frac{48{,}502\ \text{bp}}{\text{phage}} \times \frac{1.1 \times 10^{-21}\ \text{g}}{\text{bp}} \times \frac{1 \times 10^{6}\ \mu\text{g}}{\text{g}} = 66.7\ \mu\text{g DNA}$$

Therefore, a total of 66.7 μg of DNA can be recovered from 12.5×10^{12} phage. Assuming 80% recovery, this value must be multiplied by 0.8:

$$66.7\ \mu\text{g DNA} \times 0.8 = 53.4\ \mu\text{g DNA}$$

Therefore, assuming 80% recovery, 53.4 μg of DNA will be recovered from the 5 mL phage stock.

Problem 5.17 Lambda DNA is recovered from the 5 mL phage stock described in Problem 5.16. In actuality, 48.5 μg of DNA are recovered. What is the percent recovery?

Solution 5.17 % recovery = (actual yield/expected yield) $\times 100$

$$\frac{48.5\ \mu g}{66.7\ \mu g} \times 100 = 72.7\%\ \text{recovery}$$

Problem 5.18 Lambda DNA is 48,502 bp in length. What is its molecular weight?

Solution 5.18 Since the lambda genome is double-stranded DNA, the 660 daltons/bp conversion factor is used.

$$48,502\ \text{bp} \times \frac{660\ \text{daltons}}{\text{bp}} = 3.2 \times 10^{7}\ \text{daltons}$$

Problem 5.19 A plasmid cloning vector is 3250 bp in length.

(**a**) How many picomoles of vector are represented by 1 μg of purified DNA?

(**b**) How many molecules does 1 μg of this vector represent?

Solution 5.19

(**a**) The first step is to calculate the molecular weight (MW) of the vector:

$$\text{MW} = 3250\ \text{bp} \times \frac{660\ \text{g/mol}}{\text{bp}} = 2.1 \times 10^{6}\ \text{g/mol}$$

This value can then be used to calculate how many picomoles of vector are represented by 1 μg of DNA.

$$1\ \mu g\ \text{vector} \times \frac{1\ \text{mol}}{2.1 \times 10^{6}\ \text{g}} \times \frac{1 \times 10^{12}\ \text{pmol}}{1\ \text{mol}} \times \frac{1\ \text{g}}{1 \times 10^{6}\ \mu g} = \frac{1 \times 10^{12}\ \text{pmol}}{2.1 \times 10^{12}} = 0.48\ \text{pmol}$$

Therefore, 1 μg of the 3250-bp vector is equivalent to 0.48 pmol.

(**b**) To calculate the number of molecules of vector in 1 μg, we need to use the conversion factor relating base pairs and grams determined in Problem 5.15. The problem is then written as follows:

$$1\ \mu g \times \frac{1\ \text{molecule}}{3250\ \text{bp}} \times \frac{1\ \text{bp}}{1.1 \times 10^{-21}\ \text{g}} \times \frac{1\ \text{g}}{1 \times 10^{6}\ \mu g} = \frac{1\ \text{molecule}}{3.6 \times 10^{-12}} = 2.8 \times 10^{11}\ \text{molecules}$$

Therefore, there are 2.8×10^{11} molecules of the 3250-bp plasmid vector contained in 1 μg.

To calculate picomole quantities or molecule abundance for single-stranded DNA molecules, a similar approach is taken to that used in the preceding problems. However, a conversion factor of 330 g/mol should be used rather than 660 g/mol as for double-stranded DNA.

Problem 5.20 The single-stranded DNA vector M13mp18 is 7250 bases in length. How many micrograms of DNA does 1 pmol represent?

Solution 5.20 First, calculate the molecular weight of the vector:

$$7250 \text{ bases} \times \frac{330 \text{ g/mol}}{\text{base}} = 2.4 \times 10^{6} \text{ g/mol}$$

This value can then be used as a conversion factor to convert picomoles to micrograms:

$$1 \text{ pmol} \times \frac{2.4 \times 10^{6} \text{ g}}{\text{mol}} \times \frac{1 \times 10^{6} \text{ µg}}{\text{g}} \times \frac{1 \text{ mol}}{1 \times 10^{12} \text{ pmol}} = \frac{2.4 \times 10^{12} \text{ µg}}{1 \times 10^{12}} = 2.4 \text{ µg}$$

Therefore, 1 pmol of M13mp18 is equivalent to 2.4 µg DNA.

Problem 5.21 What is the molecular weight of a dephosphorylated oligonucleotide having the sequence 5′-GGACTTAGCCTTAGTATTGCCG-3′?

Solution 5.21 The oligonucleotide has four A's, five C's, six G's, and seven T's. These values are placed into the formula for calculating an oligonucleotide's molecular weight. Since the oligonucleotide is dephosphorylated, P in the equation is equal to –101.0.

$$\text{MW} = (n_{\text{A}} \times 335.2) + (n_{\text{C}} \times 311.2) + (n_{\text{G}} \times 351.2) + (n_{\text{T}} \times 326.2) + P$$

$$\text{MW} = (4 \times 335.2) + (5 \times 311.2) + (6 \times 351.2) + (7 \times 326.2) + (-101.0)$$

$$\text{MW} = (1340.8) + (1556) + (2107.2) + (2283.4) - 101.0$$

$$\text{MW} = 7186.4 \text{ daltons}$$

Estimating DNA Concentration on an Ethidium Bromide-Stained Gel

Agarose gel electrophoresis is commonly used to separate DNA fragments following restriction endonuclease digestion or PCR amplification. Fragments are detected by staining the gel with the intercalating dye, ethidium bromide, followed by visualization/photography under ultraviolet light. Ethidium bromide stains DNA in a concentration-dependent manner such that the more DNA present in a band on the gel, the more intensely it will stain. This relationship makes it possible to estimate the quantity of DNA present in a band through comparison with another band of known DNA amount. If the intensities of two bands are similar, then they contain similar amounts of DNA. Ethidium bromide stains single-stranded DNA and RNA only very poorly. These forms of nucleic acid will not give reliable quantitation by gel electrophoresis.

Problem 5.22 Five hundred nanograms (0.5 μg) of lambda DNA digested with the restriction endonuclease *Hin*dIII is loaded onto an agarose gel as a size marker. A band generated from a DNA amplification experiment has the same intensity upon staining with ethidium bromide as the 564-bp fragment from the lambda *Hin*dIII digest. What is the approximate amount of DNA in the amplified fragment?

Solution 5.22 This problem is solved by determining how much DNA is in the 564-bp fragment. Since the amplified DNA fragment has the same intensity after staining as the 564-bp fragment, the two bands contain equivalent amounts of DNA.

Phage lambda is 48,502 bp in length. The 564-bp *Hin*dIII fragment is to the total length of the phage lambda genome as its amount (in ng) is to the total amount of lambda *Hin*dIII marker run on the gel (500 ng). This allows the following relationship.

$$\frac{x \text{ ng}}{500 \text{ ng}} = \frac{564 \text{ bp}}{48{,}502 \text{ bp}}$$

$$x \text{ ng} = \frac{(500 \text{ ng})(564 \text{ bp})}{48{,}502 \text{ bp}} = 5.8 \text{ ng}$$

Therefore, there are approximately 5.8 ng of DNA in the band of the amplified DNA fragment.

Labeling Nucleic Acids with Radioisotopes

6

Introduction

Sequencing DNA, quantitating gene expression, detecting genes or recombinant clones by probe hybridization, monitoring cell replication, measuring the rate of DNA synthesis – there are a number of reasons why researchers need to label nucleic acids. Radioisotopes have found wide use as nucleic acid tags because of several unique features. They are commercially available, sparing the researcher from having to perform a great deal of up-front chemistry. They can be attached to individual nucleotides and subsequently incorporated into DNA or RNA through enzyme action. Nucleic acids labeled with radioisotopes can be detected by several methods, including liquid scintillation counting (providing information on nucleic acid quantity) and exposure to X-ray film (providing information on nucleic acid size or clone location). This section discusses the mathematics involved in the use of radioisotopes in nucleic acid research.

Units of Radioactivity: The Curie

The basic unit characterizing radioactive decay is the **curie** (**Ci**). It is defined as that quantity of radioactive substance having a decay rate of 3.7×10^{10} disintegrations per second [2.22×10^{12} **disintegrations per minute** (**dpm**)]. Most instruments used to detect radioactive decay are less than 100% efficient. Consequently, a defined number of curies of some radioactive substance will yield fewer counts than theoretically expected.

The disintegrations actually detected by an instrument, such as by liquid scintillation counter, are referred to as **counts per minute** (**cpm**).

Problem 6.1 How many dpm are associated with 1 μCi of radioactive material?

Solution 6.1 The answer is obtained by using conversion factors to cancel terms, as shown in the following equation:

$$1\ \mu\text{Ci} \times \frac{1\ \text{Ci}}{1 \times 10^{6}\ \mu\text{Ci}} \times \frac{2.22 \times 10^{12}\ \text{dpm}}{\text{Ci}} = 2.22 \times 10^{6}\ \text{dpm}$$

Therefore, 1 μCi is equivalent to 2.22×10^{6} dpm.

Problem 6.2 If the instrument used for counting radioactive material is 25% efficient at detecting disintegration events, how many cpm will 1 μCi yield?

Solution 6.2 In Problem 6.1, it was determined that 1 μCi is equivalent to 2.22×10^{6} dpm. To calculate the cpm for 1 μCi, the dpm value should be multiplied by 0.25 (25% detection efficiency):

$$2.22 \times 10^{6}\ \text{dpm} \times 0.25 = 5.55 \times 10^{5}\ \text{cpm}$$

Therefore, an instrument counting radioactive decay with 25% efficiency should detect 5.55×10^{5} cpm for a 1-μCi sample.

Estimating Plasmid Copy Number

Plasmids, autonomously replicating, circular DNA elements used for cloning of recombinant genes, exist within a cell in one to several hundred copies. In one technique for determining how many copies of a plasmid are present within an individual cell, a culture of the plasmid-carrying strain is grown in the presence of [^{3}H]-thymine or [^{3}H]-thymidine (depending on the needs of the strain). After growth to log phase, the cells are harvested and lysed, and total DNA is isolated. A sample of the isolated DNA is centrifuged in an ethidium bromide–cesium chloride gradient to separate plasmid from chromosomal DNA. Following centrifugation, the centrifuge tube is punctured at its bottom, fractions are collected onto filter discs, and the DNA is precipitated on the filters by treatment with trichloroacetic acid (TCA) and ethanol. The filters are dried and counted in a scintillation spectrometer. A graph of fraction number versus cpm might produce a profile such as shown in Figure 6.1. The smaller peak, centered at fraction number 12, represents plasmid DNA, which, in a supercoiled form, is denser than the sheared chromosomal DNA centered at fraction number 26.

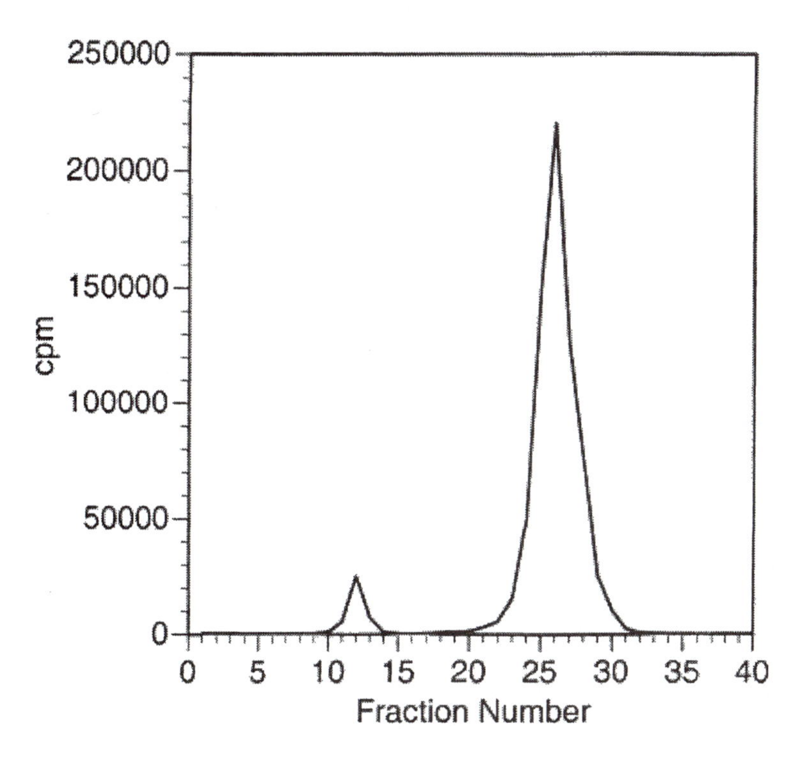

Figure 6.1 Fractions collected from an ethidium bromide–cesium chloride gradient separating plasmid from chromosomal DNA. Cells are grown in the presence of radioactive thymine or thymidine so that replicating DNA is labeled with the tritium isotope. Supercoiled plasmid DNA, centered around fraction 12, sediments at a higher density than sheared chromosomal DNA, centered around fraction 26.

Plasmid copy number is calculated using the following relationship:

$$\text{plasmid copy number} = \frac{\text{cpm of plasmid peak}}{\text{cpm of chromosome peak}} \times \frac{\text{chromosome MW}}{\text{plasmid MW}}$$

Problem 6.3 An experiment to determine the copy number of a 6000-bp plasmid in *E. coli* is performed as described earlier. The major plasmid-containing fraction is found to contain 25,000 cpm. The chromosomal peak fraction contains 220,000 cpm. What is the plasmid copy number?

Solution 6.3 The first step in solving this problem is to calculate the molecular weights of the plasmid and *E. coli* chromosome. The plasmid is 6000 bp. The *E. coli* chromosome is approximately 4.6 million bp in length. Their molecular weights are calculated using the 660-daltons/bp conversion factor.

The molecular weight of the plasmid is

$$6000 \ \text{bp} \times \frac{660 \ \text{daltons}}{\text{bp}} = 3{,}960{,}000 \ \text{daltons}$$

The molecular weight of the *E. coli* chromosome is

$$4{,}600{,}000 \ \text{bp} \times \frac{660 \ \text{daltons}}{\text{bp}} = 3.0 \times 10^{9} \ \text{daltons}$$

These values can then be incorporated into the equation (preceding this problem) for calculating copy number:

$$\frac{25{,}000 \ \text{cpm}}{220{,}000 \ \text{cpm}} \times \frac{3.0 \times 10^{9} \ \text{daltons}}{3.96 \times 10^{6} \ \text{daltons}} = \frac{7.5 \times 10^{13}}{8.7 \times 10^{11}} = 86 \ \text{plasmid copies}$$

Labeling DNA by Nick Translation

Nick translation is a technique for radioactively labeling double-stranded DNA, making it suitable as a hybridization probe for detecting specific genomic sequences. The endonuclease DNase I is used to create nicks in a DNA probe fragment. Following DNase I treatment, DNA polymerase I is used to add nucleotide residues to the free 3′-hydroxyl ends created during the DNase I nicking process. As the DNA polymerase I extends the 3′ ends, the 5′ to 3′ exonuclease activity of the enzyme removes bases from the 5′-phosphoryl terminus of the nick. The sequential addition of bases onto the 3′ end with the simultaneous removal of bases from the 5′ end of the downstream annealed strand results in translation of the nick along the DNA molecule. When performed in the presence of a radioactive deoxynucleoside triphosphate (such as [α-^{32}P] dCTP), the newly synthesized strand becomes labeled.

When measuring cpm by liquid scintillation, a blank sample should be measured containing only scintillation fluid and a filter (if a filter is used for spotting and counting of a labeling reaction). This sample is used to detect background cpm inherent in the scintillation fluid, the filter, and the environment. The background cpm should be subtracted from any sample taken from a labeling experiment.

Determining Percent Incorporation of Radioactive Label from Nick Translation
To determine the percent of radioactive label incorporated into DNA by nick translation, a sample is taken at some time after addition of the radioactive dNTP but prior to DNA polymerase I treatment. This sample represents the total cpm in the reaction. Following the nick translation reaction, another sample is withdrawn. Both samples are spotted onto filter discs. The disc containing the sample representing the reaction after nick translation is treated with TCA to precipitate DNA fragments. The TCA treatment precipitates polynucleotides onto the filter disc but allows unincorporated radioactive dNTPs to pass through. The percent incorporation is then calculated using the equation

$$\frac{\text{cpm in TCA precipitated sample}}{\text{cpm added to reaction}} \times 100 = \text{percent incorporation}$$

Problem 6.4 One microgram (1 μg) of bacteriophage lambda DNA is used for a nick translation reaction. Twenty-five μCi of labeled [α-^{32}P]dCTP are used in a 50-μL reaction. Following treatment with DNA polymerase I, a 1-μL sample is withdrawn and diluted into a total volume of 100 μL with TE buffer. Five μL of this dilution are spotted onto a glass fiber filter disc to determine total cpm in the reaction. Another 5 μL are used in a TCA precipitation procedure to determine the amount of radioisotope incorporation. The TCA-precipitated sample is collected on a glass fiber filter disc and washed with TCA and ethanol. Both filters are dried and counted by liquid scintillation. The filter disc representing the total cpm gives 19,000 cpm. The TCA-precipitated sample gives 11,600 cpm. What is the percent incorporation?

Solution 6.4 The percent incorporation is calculated using the formula described earlier. The calculation goes as follows:

$$\frac{1.16 \times 10^4 \text{ cpm}}{1.9 \times 10^4 \text{ cpm}} \times 100 = 61\% \text{ incorporation}$$

Calculating Specific Radioactivity of a Nick Translation Product

The **specific radioactivity** (or **specific activity**) of a labeled product is the amount of isotope incorporated, measured as cpm, per quantity of nucleic acid. Incorporation of isotope can be measured as TCA-precipitable cpm by the procedure described earlier. Using TCA precipitation to measure incorporation, specific radioactivity is calculated using the following formula:

$$\text{cpm}/\mu\text{g DNA} = \frac{\text{cpm in TCA precipitate}}{\text{volume used to TCA precipitate}} \times \text{dilution factor} \times \frac{\text{total reaction volume}}{\text{total } \mu\text{g DNA in reaction}}$$

Problem 6.5 In Problem 6.4, 1.16×10^4 cpm were counted as TCA-precipitable counts in a 5-µL sample taken from a 1/100 dilution of a 50-µL nick translation reaction using 1 µg of lambda DNA. What is the specific activity of the labeled product?

Solution 6.5 The incorporated radioactivity is represented by the TCA-precipitable counts. The reaction was diluted 1/100 to obtain a sample (5 µL) to TCA precipitate. The dilution factor is the inverse of the dilution. Placing these values into the preceding equation yields

$$\frac{1.16 \times 10^4 \text{ cpm}}{5 \ \mu\text{L}} \times \frac{100 \ \mu\text{L}}{1 \ \mu\text{L}} \times \frac{50 \ \mu\text{L}}{1 \ \mu\text{g DNA}} = \frac{5.8 \times 10^7 \text{ cpm}}{5 \ \mu\text{g}} = 1.16 \times 10^7 \text{ cpm}/\mu\text{g}$$

Therefore, the nick translation product has a specific activity of 1.16×10^7 cpm/µg.

Random Primer Labeling of DNA

An oligonucleotide six bases long is called a **hexamer** or **hexanucleotide**. If synthesis of a hexamer is carried out in such a way that A, C, G, and T nucleotides have an equal probability of coupling at each base addition step, then the resulting product will be a mixture of oligonucleotides having many different random sequences. Such an oligonucleotide is called a **random hexamer**. For any six contiguous nucleotides on DNA derived from a biological source, there should be a hexamer within the random hexamer pool that will be complementary. Under the proper conditions of temperature and salt, the annealed hexamer can then serve as a primer for in vitro DNA synthesis. By using a radioactively labeled dNTP (such as

[α-^{32}P]dCTP) in the DNA synthesis reaction, the synthesis products become radioactively labeled. The labeled DNA can then be used as a probe to identify specific sequences by any of a number of gene detection protocols using hybridization.

Several mathematical calculations are necessary to assess the quality of the reaction, and these are outlined in the following sections.

Random Primer Labeling – Percent Incorporation

The percent incorporation for a random-primer-labeling experiment is determined in the same manner as that used in the nick translation procedure. It is calculated by dividing the cpm incorporated into DNA (as TCA-precipitable cpm) by the total cpm in the reaction and then multiplying this value by 100:

$$\frac{\text{cpm incorporated into DNA}}{\text{total cpm in reaction}} \times 100 = \% \text{ incorporation}$$

Problem 6.6 In a random-primer-labeling experiment, random hexamers, 90 ng of denatured DNA template, DNA polymerase Klenow enzyme, enzyme buffer, 100 μCi [α-^{32}P]dCTP (2000 Ci/mmol), and a mixture of the three nonisotopically labeled dNTPs (at molar excess to labeled dCTP) are incubated in a 70-μL reaction at room temperature for 75 minutes to allow DNA synthesis by extension of the annealed primers. The reaction is terminated by heating. A 1-μL aliquot of the reaction is diluted into 99 μL of TE buffer. Five μL of this dilution are spotted onto a filter disc, dried, and counted by liquid scintillation to measure total cpm in the reaction. Another 5 μL of the 1/100 dilution are used in a TCA precipitation procedure to measure the amount of label incorporated into polynucleotides. The filter disc used to measure total cpm gives 6.2×10^4 cpm. The TCA-precipitated sample gives 5.1×10^4 cpm. What is the percent incorporation?

Solution 6.6 Using the formula for calculating % incorporation given earlier, the solution is calculated as follows:

$$\frac{5.1 \times 10^4 \text{ cpm}}{6.2 \times 10^4 \text{ cpm}} \times 100 = 82\% \text{ incorporation}$$

Therefore, the reaction achieved 82% incorporation of the labeled base.

Random Primer Labeling – Calculating Theoretical Yield

Nick translation replaces existing DNA with radioactively labeled strands with no net addition of DNA quantity. Random primer labeling, on the other hand, actually results in the production of completely new and additional DNA strands. To calculate theoretical yield, it is assumed that the reaction achieves 100% incorporation of the labeled dNTP. In the random-primer-labeling experiment, the unlabeled dNTPs are placed into the reaction in molar excess to the labeled dNTP. The unlabeled dNTPs can be at a concentration 10- to 100-fold higher than that of the radioactively labeled dNTP.

Theoretical yield is calculated using the following formula:

$$\text{grams DNA} = \frac{\mu\text{Ci dNTP added to reaction} \times 4 \text{ dNTPs} \times \dfrac{330 \text{ g/mol}}{\text{dNTP}}}{\text{dNTP specific activity in } \mu\text{Ci/mol}}$$

In this formula, all four dNTPs are assumed to have the same concentration as that of the labeled dNTP, since it is the labeled dNTP that is in limiting amount.

In a typical random-primer-labeling experiment, nanogram amounts of DNA are synthesized. The preceding equation can be converted to a form that more readily yields an answer in nanograms of DNA. The average molecular weight of a nucleotide (dNTP) is 330 g/mol. This is equivalent to 330 ng/nmol as shown in the following equation:

$$\frac{330 \text{ g}}{\text{mol}} \times \frac{1 \times 10^9 \text{ ng}}{\text{g}} \times \frac{1 \text{ mol}}{1 \times 10^9 \text{ nmol}} = \frac{330 \text{ ng}}{\text{nmol}}$$

The specific activity of a radioactive dNTP, such as [α-^{32}P]dCTP, from most suppliers, is expressed in units of Ci/mmol. This unit is equivalent to μCi/nmol:

$$\frac{\text{Ci}}{\text{mmol}} \times \frac{1 \times 10^6 \ \mu\text{Ci}}{\text{Ci}} \times \frac{\text{mmol}}{1 \times 10^6 \text{ nmol}} = \frac{\mu\text{Ci}}{\text{nmol}}$$

Placing these new conversion factors into the equation for calculating theoretical yield gives

$$\text{ng DNA} = \frac{\mu\text{Ci dNTP added} \times 4 \ \text{dNTPs} \times \dfrac{330 \ \text{ng/nmol}}{\text{dNTP}}}{\text{dNTP specific activity in } \mu\text{Ci/nmol}}$$

Problem 6.7 In Problem 6.6, 100 µCi of [α-^{32}P]dCTP having a specific activity of 2000 Ci/mmole was used in a random-primer-labeling experiment. Assuming 100% efficient primer extension, what is the theoretical yield?

Solution 6.7 As shown earlier, a dNTP-specific activity of 2000 Ci/mmol is equivalent to 2000 µCi/nmol. The equation for calculating yield is then as follows:

$$\text{ng DNA} = \frac{100 \ \mu\text{Ci} \times 4 \ \text{dNTPs} \times \dfrac{330 \ \text{ng/nmol}}{\text{dNTP}}}{2000 \ \mu\text{Ci dNTP/nmol}}$$

$$= \frac{132{,}000 \ \text{ng}}{2000} = 66 \ \text{ng DNA theoretical yield}$$

Therefore, in the random-primer-labeling experiment described in Problem 6.6, 66 ng of DNA should be synthesized if the reaction is 100% efficient.

Random Primer Labeling – Calculating Actual Yield

The actual yield of DNA synthesized by random primer labeling is calculated by multiplying the theoretical yield by the % incorporation (expressed as a decimal):

$$\text{theoretical yield} \times \% \ \text{incorporation} = \text{ng DNA synthesized}$$

Problem 6.8 In Problem 6.6, it was calculated that the described reaction achieved 82% incorporation of the labeled dNTP. In Problem 6.7, it was calculated that the random-primer-labeling experiment described in Problem 6.6 would theoretically yield 66 ng of newly synthesized DNA. What is the actual yield?

Solution 6.8 The actual yield is obtained by multiplying the % incorporation by the theoretical yield. Expressed as a decimal, 82% (the percent incorporation) is 0.82. Using this value for calculating actual yield gives the following equation:

$$0.82 \times 66 \text{ ng DNA} = 54.1 \text{ ng DNA}$$

Therefore, under the conditions of the random-primer-labeling experiment described in Problem 6.6, an actual yield of 54.1 ng of newly synthesized DNA is obtained.

Random Primer Labeling – Calculating Specific Activity of the Product

The specific activity of a DNA product generated by random primer labeling is expressed as cpm/μg DNA. It is calculated using the following formula:

$$\frac{\dfrac{\text{cpm incorporated}}{\text{mL used to measure incorporation}} \times \text{dilution factor} \times \text{reaction volume}}{(\text{ng DNA actual yield} + \text{ng input DNA}) \times \dfrac{0.001\ \mu\text{g}}{\text{ng}}} = \text{cpm}/\mu\text{g}$$

Problem 6.9 Problem 6.6 describes a random-primer-labeling experiment in which 90 ng of denatured template DNA are used in a 70-μL reaction. To measure incorporation of the labeled dNTP, 5 μL of a 1/100 dilution taken from the 70-μL reaction are TCA precipitated on a filter and counted. This 5-μL sample gives 5.1×10^4 cpm. In Problem 6.8, an actual yield of 54.1 ng of DNA was calculated for this labeling experiment. What is the specific activity of the DNA synthesis product?

Solution 6.9 Substituting the given values into the preceding equation yields the following result. Note the use of the dilution factor when calculating the cpm incorporated.

$$\frac{\dfrac{5.1 \times 10^4 \text{ cpm}}{5\ \mu\text{L}} \times \dfrac{100\ \mu\text{L}}{1\ \mu\text{L}} \times 70\ \mu\text{L}}{(54.1 \text{ ng DNA} + 90 \text{ ng DNA}) \times \dfrac{0.001\ \mu\text{g}}{\text{ng}}} = \frac{7.1 \times 10^7 \text{ cpm}}{0.144\ \mu\text{g}} = 4.9 \times 10^8 \text{ cpm}/\mu\text{g DNA}$$

Therefore, the newly synthesized DNA in this random-primer-labeling experiment has a specific activity of 4.9×10^8 cpm/μg.

Labeling 3′ Termini with Terminal Transferase

The enzyme **terminal deoxynucleotidyl transferase** adds mononucleotides onto 3′ hydroxyl termini of either single- or double-stranded DNA. The mononucleotides are donated by a dNTP. Addition of each mononucleotide releases inorganic phosphate. In a procedure referred to as **homopolymeric tailing** (or simply "**tailing**"), terminal transferase is used to add a series of bases onto the 3′ end of a DNA fragment. The reaction can also be manipulated so that only one mononucleotide is added onto the 3′ end. This is accomplished by supplying the reaction exclusively with either dideoxynucleoside 5′-triphosphate (ddNTP) or cordycepin-5′-triphosphate. Both of these analogs lack a free 3′ hydroxyl group. Once one residue has been added, polymerization stops, since the added base lacks the 3′ hydroxyl required for coupling of a subsequent mononucleotide. By either approach, use of a labeled dNTP donor provides another method of preparing a labeled probe suitable for use in gene detection hybridization assays.

3′ -End Labeling with Terminal Transferase – Percent Incorporation

As with the other labeling methods described in this chapter, TCA precipitation (described earlier) can be used to measure incorporation of a labeled dNTP. Percent incorporation is calculated using the following equation:

$$\frac{\text{cpm incorporated}}{\text{total cpm}} \times 100 = \%\ \text{incorporation}$$

Problem 6.10 A 3′-end-labeling experiment is performed in which 60 μg of DNA fragment, terminal transferase, buffer, and 100 μCi of [α-^{32}P] cordycepin-5′-triphosphate (2000 Ci/mmol) are combined in a 50-μL reaction. The mixture is incubated at 37°C for 20 minutes and stopped by heating at 72°C for 10 minutes. One μL of the reaction is diluted into 99 μL of TE buffer. Five μL of this dilution are spotted onto a filter; the filter is dried and then counted in a liquid scintillation counter. This sample, representing total cpm, gives 90,000 cpm. Another 5 μL of the 1/100 dilution are treated with TCA to measure the amount of label incorporated into DNA. The filter from the TCA procedure gives 76,500 cpm. What is the percent incorporation?

Solution 6.10 The total cpm was measured to be 90,000 cpm. By TCA precipitation, it was found that 76,500 cpm are incorporated into DNA. Using these values and the equation shown earlier, the calculation goes as follows:

$$\frac{76,500 \text{ cpm}}{90,000 \text{ cpm}} \times 100 = 85\% \text{ incorporation}$$

Therefore, in this 3′-end-labeling experiment, 85% of the label was incorporated into DNA.

3′- End Labeling with Terminal Transferase – Specific Activity of the Product

The specific activity of the 3′-end-labeled product, expressed as cpm/μg DNA, is given by the equation

$$\text{cpm/}\mu\text{g} = \frac{\%\text{ incorp.} \times \dfrac{\text{total cpm}}{\mu\text{L used to measure total cpm}} \times \text{dilution factor} \times \text{reaction volume}}{\mu\text{g of DNA fragment in reaction}}$$

Problem 6.11 In Problem 6.10, a 3′-end-labeling reaction is described in which 60 μg of DNA fragment are labeled with [α-^{32}P] cordycepin-5′-triphosphate in a 50-μL reaction. A 1-μL aliquot of the reaction is diluted 1 μL/100 μL, and 5 μL are counted, giving 90,000 cpm. In Problem 6.10, it was shown that 85% of the radioactive label was incorporated into DNA. What is the specific activity of the product?

Solution 6.11 Using the preceding equation yields

$$\frac{0.85 \times \dfrac{90,000 \text{ cpm}}{5\ \mu\text{L}} \times \dfrac{100\ \mu\text{L}}{1\ \mu\text{L}} \times 50\ \mu\text{L}}{60\ \mu\text{g DNA}} = \frac{7.65 \times 10^{7} \text{ cpm}}{60\ \mu\text{g DNA}} = 1.3 \times 10^{6} \text{ cpm/}\mu\text{g DNA}$$

Therefore, the product of this 3′-end-labeling experiment has a specific activity of 1.3×10^{6} cpm/μg.

cDNA Synthesis

RNA is a fragile molecule and is easily degraded when released from a cell during an extraction protocol. This can make the study of gene expression a challenging task. One method to make the study of RNA more accessible has been to copy RNA into DNA. Double-stranded DNA is far more stable and amenable to manipulation by recombinant DNA techniques than is RNA.

The retroviral enzyme **reverse transcriptase** has the unique ability to synthesize a complementary DNA strand from an RNA template. DNA produced in this manner is termed **cDNA** (for complementary DNA). Once made, a DNA polymerase enzyme can be used in a second step to synthesize a complete double-stranded cDNA molecule using the first cDNA strand as template. The cloning and sequencing of cDNAs have provided molecular biologists with valuable insights into gene structure and function.

First Strand cDNA Synthesis

Since most eukaryotic mRNAs can be isolated having intact poly (A) tails, an oligo(dT) primer designed to anneal to homopolymer tails can be used to initiate synthesis of the first cDNA strand. When long RNA molecules are to be reverse transcribed, however, using random hexanucleotide primers is a more favored approach. Other reaction components include reverse transcriptase, a ribonuclease inhibitor, the four dNTPs, dithiothreitol (a reducing agent), and a reaction buffer. RNA is usually added to the reaction at a concentration of roughly 0.1 µg/µL. When all reaction components have been combined, a small aliquot of the reaction is transferred to a separate tube containing a small amount of [α-^{32}P]dCTP to monitor first strand synthesis. The original reaction mixture (with no radioisotope) is carried through synthesis of the second strand.

Incorporation of label into polynucleotide can be followed by TCA precipitation, as described previously for other labeling reactions (see page 113). Calculating percent incorporation of label is performed as described previously in this chapter. The purpose of assessing the first strand cDNA synthesis reaction is to determine the quantity of cDNA synthesized and the efficiency with which mRNA was converted to cDNA. These calculations require the following determinations.

a) the moles of dNTP per reaction given by the equation

$$\text{moles of dNTP/reaction} = (\text{molar concentration of dNTP}) \times (\text{reaction volume})$$

b) The percent incorporation of labeled dNTP given by the equation

$$\% \text{ incorporation} = \frac{\text{TCA precipitable cpm}}{\text{total cpm}} \times 100$$

c) Moles of dNTP incorporated into polynucleotide given by the equation

$$\text{moles dNTP incorporated} = \frac{\text{moles dNTP}}{\mu\text{L}} \times \text{reaction volume} \times \frac{\% \text{ incorporation}}{100}$$

d) Grams of first strand cDNA synthesized given by the equation

$$\text{grams cDNA synthesized} = \text{moles dNTP incorporated} \times \frac{330 \text{ grams}}{\text{moles dNTP}}$$

e) The number of grams of first strand cDNA carried through to second strand synthesis given by the equation

$$\text{grams cDNA available for second strand synthesis} =$$

$$\text{grams first strand cDNA} \times \frac{\text{reaction volume after sample removed for labeling}}{\text{initial reaction volume}}$$

Problem 6.12 A first strand cDNA synthesis experiment is performed in which 5 μg of poly (A) mRNA are combined with reverse transcriptase, 2 m*M* each dNTP, and other necessary reagents in a total volume of 55 μL. After all components have been mixed, 5 μL of the reaction are transferred to a different tube containing 0.5 μL of [α-^{32}P]dCTP (1000 Ci/mmol; 10 μCi/μL). In the small side reaction with the [α-^{32}P]dCTP tracer, total cpm and incorporation of label are measured by TCA precipitation in the following manner. Two μL of the side reaction are diluted into 100 μL of TE buffer. Five μL of this dilution are spotted onto a filter and counted (to measure total cpm). Another 5 μL of the 2/100 dilution of the side reaction are precipitated with TCA and counted. The sample prepared to measure total cpm gives 220,000 cpm. The TCA-precipitated sample gives 2000 cpm. (**a**) How much cDNA was synthesized in this reaction? (**b**) How much of the input mRNA was converted to cDNA?

Solution 6.12(a) This problem can be solved using the equations shown earlier. The first step is to determine the concentration of dNTPs (expressed as nmol/μL). The reaction contains 2 m*M* dATP, 2 m*M* dCTP, 2m*M* dGTP, and 2 m*M* dTTP, or a total dNTP concentration of 8m*M*. An 8 m*M* dNTP concentration is equivalent to 8 mmol/L. This can be converted to nmol/μL using the following relationship:

$$\frac{8 \text{ mmol dNTP}}{\text{L}} \times \frac{1 \text{ L}}{1 \times 10^{6} \ \mu\text{L}} \times \frac{1 \times 10^{6} \text{ nmol}}{\text{mmol}} = \frac{8 \text{ nmol dNTP}}{\mu\text{L}}$$

The percent incorporation of [α-^{32}P]dCTP can be calculated using the data provided:

$$\frac{2000 \text{ incorporated cpm}}{220{,}000 \text{ total cpm}} \times 100 = 0.9\% \text{ incorporation}$$

The number of nanomoles of dNTP incorporated into cDNA can now be calculated. To perform this step, the concentration of dNTPs in the reaction (expressed in nmol dNTP/μL), the total volume of the reaction prior to sample aliquoting into a separate tracer reaction, and the percent incorporation (expressed as a decimal) are used as follows:

$$\frac{8 \text{ nmol dNTP}}{\mu\text{L}} \times 55 \ \mu\text{L} \times 0.009 = 4 \text{ nmol dNTP incorporated}$$

Using the number of nanomoles of dNTP incorporated into nucleic acid and the average molecular weight of a dNTP (330 g/mol), the amount of cDNA synthesized can be calculated by multiplying these two values. Conversion factors must be applied in this calculation, since we have determined nanomoles of incorporated dNTP and the molecular weight of a dNTP is given in terms of grams/mole. The calculation goes as follows:

$$4 \text{ nmol dNTP} \times \frac{330 \text{ g dNTP}}{\text{mol}} \times \frac{1 \times 10^{9} \text{ ng}}{\text{g}} \times \frac{\text{mol}}{1 \times 10^{9} \text{ nmol}} = 1320 \text{ ng cDNA}$$

Therefore, in this reaction, 1320 ng of cDNA were synthesized.

Solution 6.12(b) The percent mRNA converted to cDNA is calculated by dividing the amount of cDNA synthesized by the amount of mRNA put into the reaction. The quotient is then multiplied by 100. In this example, 1320 ng of cDNA were synthesized. Five μg of mRNA were put into the reaction. Since these nucleic acid quantities are expressed in different units, they first must be brought to equivalent units. Converting μg mRNA to ng mRNA is performed as follows:

$$5 \ \mu\text{g mRNA} \times \frac{1000 \text{ ng}}{\mu\text{g}} = 5000 \text{ ng mRNA}$$

Percent mRNA converted to cDNA can now be calculated:

$$\frac{1320 \text{ ng cDNA synthesized}}{5000 \text{ ng mRNA in reaction}} \times 100 = 26.4\% \text{ mRNA converted to cDNA}$$

Therefore, 26.4% of the mRNA put into the reaction was converted to cDNA.

Problem 6.13 How much of the cDNA synthesized in Problem 6.12 will be carried into the second strand synthesis reaction?

Solution 6.13 In Problem 6.12, 5 μL of a 55-μL reaction were removed to monitor incorporation of [α-^{32}P]dCTP. Therefore, 50 μL of 55 μL remained in the reaction. Multiplying this ratio by the amount of cDNA synthesized (1320 ng) gives the following result:

$$1320 \ \text{ng} \times \frac{50 \ \mu\text{L}}{55 \ \mu\text{L}} = 1200 \ \text{ng cDNA carried to second strand synthesis}$$

Problem 6.14 For the cDNA synthesis reaction described in Problem 6.12, how many nanomoles of dNTP were incorporated into cDNA in the main reaction?

Solution 6.14 In Problem 6.12, a total of 4 nmol of dNTP were incorporated into nucleic acid. Multiplying this amount by the volume fraction left in the main reaction will give the number of nanomoles of dNTP incorporated into cDNA that will be carried into the second strand synthesis reaction:

$$4 \ \text{nmol dNTP incorporated} \times \frac{50 \ \mu\text{L}}{55 \ \mu\text{L}} = 3.6 \ \text{nmol dNTP incorporated}$$

Therefore, 3.6 nmol of dNTP were incorporated into cDNA in the main reaction.

Second Strand cDNA Synthesis

Synthesis of the cDNA second strand can be accomplished by a number of different methods. As a first step, the RNA in the RNA–cDNA hybrid formed during the first strand synthesis reaction can be removed by treatment with RNase A. Priming of second strand synthesis can then be performed using random primers, a gene-specific primer, or the hairpin loop structure typically found at the 3′-end of the first strand cDNA. Alternatively, RNase H can be used to nick the RNA strand of the RNA–cDNA hybrid molecule. DNA polymerase then utilizes the nicked RNA as primer to replace the RNA strand in a reaction similar to nick translation.

To perform second strand cDNA synthesis, the first strand reaction is diluted directly into a new mixture carrying all the components necessary for successful DNA replication. Typically, no new dNTPs are added. They are carried into the second strand reaction from the first strand reaction. As with first strand cDNA synthesis, a separate tracer reaction including [α-^{32}P]dCTP can be used to monitor incorporation of dNTPs and formation of the DNA second strand.

Several equations are used to determine the quality of second strand cDNA synthesis. (a) Percent incorporation of label into the second strand is calculated using the following formula:

$$\frac{\text{TCA precipitable cpm}}{\text{total cpm}} \times 100 = \%\text{ incorporation into second strand}$$

(b) Nanomoles of dNTP incorporated into cDNA is given by the following equation:

$$\text{nmol dNTP incorporated} = [\text{nmol dNTP}/\mu\text{L} \times \text{reaction volume} - \text{nmol dNTP incorporated in first cDNA strand}] \times \%\text{ incorporation}$$

(In this equation, % incorporation should be expressed as a decimal.)

(c) The amount of second strand cDNA synthesized (in ng) is given by the following equation:

$$\text{ng second strand cDNA} = \text{nmol dNTP incorporated} \times \frac{330\ \text{ng dNTP}}{\text{nmol}}$$

(d) The percent of first strand cDNA converted to double-stranded cDNA is given by the following equation:

$$\%\text{ conversion to double-stranded cDNA} = \frac{\text{ng second strand cDNA}}{\text{ng first strand cDNA}} \times 100$$

Problem 6.15 The 50-μL first strand cDNA synthesis reaction described in Problem 6.12 is diluted into a 250-μL second strand synthesis reaction containing all the necessary reagents. To prepare a tracer reaction, 10 μL of the 250-μL main reaction are transferred to a separate tube containing 5 μCi [α-^{32}P]dCTP. Following incubation of the reactions, 90 μL of an EDTA solution are added to the tracer reaction. A heating step is used to stop the reaction. Two μL of the 100-μL tracer reaction are spotted on a filter and counted. This sample gives 90,000 cpm. Another 2-μL sample from the diluted tracer reaction is treated with TCA to measure incorporation of [α-^{32}P]dCTP. This sample gives 720 cpm. **(a)** How many nanograms of second strand cDNA were synthesized? **(b)** What percentage of the first strand was converted into second strand cDNA?

Solution 6.15(a) The first step in solving this problem is to calculate the percent incorporation of [α-^{32}P]dCTP:

$$\frac{720\ \text{TCA precipitable cpm}}{90{,}000\ \text{total cpm}} \times 100 = 0.8\%\ \text{incorporation}$$

The nanomoles of dNTP incorporated can now be calculated. The original first strand cDNA synthesis reaction had a dNTP concentration of 8 nmol/μL (see Solution 6.12a). The 50-μL first strand main reaction was diluted into a total volume of 250 μL. Assuming no incorporation of dNTP in the first strand reaction, the new dNTP concentration in the second strand reaction is

$$\frac{8\ \text{nmol dNTP}}{\mu\text{L}} \times \frac{50\ \mu\text{L}}{250\ \mu\text{L}} = \frac{1.6\ \text{nmol dNTP}}{\mu\text{L}}$$

The first strand of cDNA incorporated 3.6 nmol of dNTP in the main reaction (Problem 6.14). Since this amount of dNTP is no longer available in the second strand synthesis reaction for incorporation into nucleic acid, it must be subtracted from the total dNTP concentration. Using the % incorporation calculated earlier, the amount of dNTP incorporated can be calculated from the following equation:

$$\left[\left(\frac{1.6 \text{ nmol}}{\mu\text{L}} \times 250 \;\; \mu\text{L}\right) - 3.6 \;\; \text{nmol dNTP incorporarated in first strand}\right]$$
$$\times 0.008 = 3.2 \;\; \text{nmol dNTP incorporated}$$

Therefore, 3.2 nmol of dNTP were incorporated into cDNA during synthesis of the second strand.

The amount of cDNA produced during second strand synthesis can now be calculated by multiplying the nanomoles of dNTP incorporated by the average molecular weight of a dNTP (330 ng/nmol):

$$3.2 \;\; \text{nmol dNTP} \times \frac{330 \;\; \text{ng}}{\text{nmol}} = 1056 \;\; \text{ng second strand cDNA}$$

Therefore, 1056 ng of cDNA were synthesized in the second strand cDNA synthesis reaction.

Solution 6.15(b) In Problem 6.13, it was calculated that 1200 ng of cDNA synthesized in the first strand synthesis reaction were carried into the second strand cDNA synthesis reaction. The percent conversion of input nucleic acid to second strand cDNA is calculated by dividing the amount of second strand cDNA synthesized by the amount of first strand cDNA carried into the second strand reaction and multiplying this value by 100:

$$\frac{1056 \text{ ng second strand cDNA}}{1200 \text{ ng first strand cDNA}} \times 100 = 88\% \text{ conversion to double - stranded cDNA}$$

Therefore, 88% of the input first strand cDNA was converted to double-stranded cDNA in the second strand synthesis reaction.

Homopolymeric Tailing

Homopolymeric tailing (or simply "**tailing**") is the process by which the enzyme **terminal transferase** is used to add a series of residues (all of the same base) onto the 3′ termini of double-stranded DNA. It has found particular use in facilitating the cloning of cDNA into plasmid cloning vectors. In this technique, a double-stranded cDNA fragment is tailed with a series of either dC or dG residues onto the two 3′ termini. The vector into which the cDNA fragment is to be cloned is tailed with the complementary base (dC if dG was used to tail the cDNA or dG if dC was used to tail the cDNA). By tailing the cDNA and the vector with complementary bases, effective annealing between the two DNAs is promoted. Ligation of the cDNA fragment to the vector is then more efficient:

It is possible to determine the number of residues that terminal transferase has added onto a 3′ end by determining the number of picomoles of 3′ ends available as substrate and the concentration (in picomoles) of dG (or dC) incorporated into DNA. As to the first part, the number of picomoles of 3′ ends is calculated by using the DNA substrate's size in the following formula:

$$\text{pmoles 3' ends} = \frac{\mu\text{g DNA substrate} \times \dfrac{1\text{ g}}{1 \times 10^{6}\ \mu\text{g}}}{\dfrac{\text{number of bp}}{\text{molecule}} \times \dfrac{660\text{ g/mol}}{\text{bp}}} \times \frac{1 \times 10^{12}\text{ pmol}}{\text{mol}} \times \frac{\text{number of 3' ends}}{\text{molecule}}$$

Problem 6.16 If 2.5 μg of a 3200-bp plasmid cloning vector are cut once with a restriction enzyme, how many picomoles of 3′ ends are available as substrate for terminal transferase?

Solution 6.16 A plasmid is a circular double-stranded DNA molecule. If it is cut with a restriction endonuclease at one site, it will become linearized. As a linear molecule, it will have two 3′ ends. The equation for calculating the amount of 3′ ends (see earlier) becomes

$$\text{pmol 3' ends} = \frac{2.5\ \mu\text{g DNA} \times \dfrac{1\ \text{g}}{1 \times 10^6\ \mu\text{g}}}{\dfrac{3200\ \text{bp}}{\text{molecule}} \times \dfrac{660\ \text{g/mol}}{\text{bp}}} \times \frac{1 \times 10^{12}\ \text{pmol}}{\text{mol}} \times \frac{2\ \text{3' ends}}{\text{molecule}}$$

$$= \frac{5 \times 10^6\ \text{pmol 3' ends}}{2.1 \times 10^6} = 2.4\ \text{pmol 3' ends}$$

Therefore, 2.5 μg of a linear DNA fragment 3200 bp in length are equivalent to 2.4 pmol of 3′ ends.

To estimate the number of dG residues added onto a 3′ end of a DNA fragment, it is necessary to have a radioactive molecule present in the reaction that can be detected and quantitated. The molecule used to monitor the progress of the reaction should be a substrate of the enzyme. In a dG tailing protocol, therefore, tritiated dGTP (^{3}H-dGTP) can serve this purpose. The next step in calculating the number of residues added to a 3′ end is to calculate the specific activity of the dNTP. Specific activity can be expressed as cpm/pmol dNTP. The general formula for calculating the specific activity of tailed DNA is

$$\text{cpm/pmol dNTP} = \frac{\text{total cpm in counted sample}}{\text{pmol dNTP}}$$

To perform this calculation, the number of picomoles of dNTP in the reaction must be determined. Let's assume that the concentration of the starting dNTP material, as supplied by the manufacturer, is expressed in a micromolar (μ*M*) amount. The micromolar concentration of dNTP in the terminal transferase tailing reaction mix is determined by the following formula:

$$\mu M\ \text{dNTP} = \frac{(\mu M\ \text{concentration of dNTP stock}) \times (\text{volume dNTP added})}{\text{total reaction volume}}$$

Problem 6.17 A 3200-bp vector is being tailed with dG residues using dGTP and terminal transferase. The stock solution of dGTP being used as substrate has a concentration of 200 μ*M*. Four μL of this stock solution are added to a 100-μL reaction (final volume). What is the micromolar concentration of dGTP in the reaction?

Solution 6.17 Using the equation just given yields the following relationship:

$$\mu\text{M dGTP} = \frac{(200\ \mu\text{M}) \times (4\ \mu\text{L})}{100\ \mu\text{L}} = 8\ \mu\text{M}$$

Therefore, dGTP in the reaction mix has a concentration of 8 μM.

The dGTP in the reaction is usually used in large molar excess to the ^{3}H-dGTP added. The added ^{3}H-dGTP, therefore, does not contribute significantly to the overall dGTP concentration. When an aliquot of the radioactive dGTP is added to the terminal transferase tailing reaction, the total cpm can be determined by spotting a small amount of the reaction mix on a filter, allowing the filter to dry, and then counting by liquid scintillation. The number of picomoles of dGTP spotted on the filter is given by the following equation:

$$\text{pmol dNTP} = \frac{\text{mmol dNTP}}{\text{L}} \times \frac{\text{L}}{1 \times 10^6\ \mu\text{L}} \times \frac{1 \times 10^6\ \text{pmol}}{\mu\text{mol}} \times \mu\text{L spotted on filter}$$

It must be remembered that μM is equivalent to micromoles/liter (see Chapter 2). This gives us the first term of this equation. For example, an 8 μM solution of dGTP is equivalent to 8 μmol dGTP/L.

Problem 6.18 The terminal transferase tailing reaction just described has a dGTP concentration of 8 μM. Five μCi of ^{3}H-dGTP are added to the reaction as a reporter molecule. Two microliters of the reaction mix are then spotted on a filter and counted by liquid scintillation to determine total cpm in the reaction. How many picomoles of dGTP are on the filter?

Solution 6.18 The ^{3}H-dGTP added to the reaction is such a small molar amount that it does not significantly change the overall dGTP concentration of 8 μM. The earlier equation yields the following relationship:

$$\text{pmol dGTP} = \frac{8\ \mu\text{mol}}{\text{L}} \times \frac{\text{L}}{1 \times 10^6\ \mu\text{L}} \times \frac{1 \times 10^6\ \text{pmol}}{\mu\text{mol}} \times 2\ \mu\text{L} = 16\ \text{pmol dGTP}$$

Therefore, the 2 μL of reaction spotted on the filter to measure total cpm contains 16 pmol of dGTP.

A specific activity can be calculated by measuring the cpm of the spotted material and by use of the following formula:

$$\text{specific activity (in cpm/pmol)} = \frac{\text{spotted filter cpm}}{\text{pmol of dNTP in spotted material}}$$

Problem 6.19 The 2-μL sample spotted on the filter (referred to in Problem 6.18) was counted by liquid scintillation and gave 90,000 cpm. What is the specific activity, in cpm/pmol dNTP, of the spotted sample?

Solution 6.19 In Problem 6.18 , it was calculated that a 2-μL sample of the tailing reaction contains 16 pmol of dGTP. Placing this value in the preceding equation gives the following relationship:

$$\text{specific activity} = \frac{90{,}000\ \text{cpm}}{16\ \text{pmol dGTP}} = 5625\ \text{cpm/pmol dGTP}$$

Therefore, the tailing reaction has a specific activity of 5625 cpm/pmol dGTP.

Incorporation of label into a DNA strand can be determined by TCA precipitation. Knowing the specific activity of the tailing dNTP allows for the determination of the total number of pmoles of base incorporated into a homopolymeric tail. This calculation is done using the following formula:

$$\text{total pmol dNTP incorporated} = \frac{\dfrac{\text{TCA - precipitable cpm}}{\mu\text{L used for TCA precipitation}} \times \text{total}\ \ \mu\text{L in reaction}}{\text{specific activity of dNTP (in cpm/pmol dNTP)}}$$

Problem 6.20 Following incubation with terminal transferase, 5 μL of the reaction mixture are treated with TCA and ethanol to measure incorporation of dGTP into polynucleotide. The filter from this assay gives 14,175 cpm by liquid scintillation. How many picomoles of dGTP were incorporated into a homopolymeric tail?

Solution 6.20 It was calculated in Problem 6.19 that the dGTP in the tailing reaction has a specific activity of 5625 cpm/pmol. The reaction volume is 100 μL. Placing these values into the preceding equation written yields the following relationship:

$$\text{total pmol dGTP inorporated} = \frac{\dfrac{14{,}175 \text{ TCA-precipitable cpm}}{5 \ \mu\text{L}} \times 100 \ \mu\text{L}}{5625 \ \text{cpm/pmol dGTP}} = 50.4 \ \text{pmol dGTP}$$

Therefore, a total of 50.4 picomoles of dGTP were incorporated into homopolymeric tails in this reaction.

The number of nucleotide residues added per 3′ end can be calculated using the following formula:

$$\text{nucleotide residues/3'end} = \frac{\text{pmol dNTP incorporated}}{\text{pmol 3' ends}}$$

It should be remembered that the term “picomoles” represents a number, a quantity. Picomoles of different molecules are related by Avogadro’s number (see Chapter 2). One mole of anything is equivalent to 6.023×10^{23} units. One picomole of anything, therefore, consists of 6.023×10^{11} units, as shown here:

$$\frac{6.023 \times 10^{23} \ \text{molecules}}{\text{mol}} \times \frac{\text{mole}}{1 \times 10^{12} \ \text{pmol}} = \frac{6.023 \times 10^{11} \ \text{molecules}}{\text{pmol}}$$

Let’s assume we find that 20 pmol dGTP have been incorporated into polynucleotides in a terminal transferase tailing reaction. Assume further that there are 2 pmol 3′ ends onto which tails can be added. Twenty picomoles of dGTP is equivalent to 1.2×10^{13} molecules of dGTP, as shown in this equation:

$$\frac{6.023 \times 10^{11} \ \text{dGTP molecules}}{\text{pmol dGTP}} \times 20 \ \text{pmol dGTP} = 1.2 \times 10^{13} \ \text{molecules dGTP}$$

Two picomoles of 3′ ends are equivalent to 1.2×10^{12} 3′ ends:

$$\frac{6.023 \times 10^{11} \ \text{3' ends}}{\text{pmol 3' ends}} \times 2 \ \text{pmol 3' ends} = 1.2 \times 10^{12} \ \text{3' ends}$$

Whether the number of residues added to the available 3′ ends is calculated using picomoles or using molecules, the ratios are the same, in this case 10:

$$\text{dG residues added} = \frac{20 \text{ pmol dGTP incorporated}}{2 \text{ pmol 3' ends}} = 10$$

or

$$\text{dG residues added} = \frac{1.2 \times 10^{13} \text{ molecules dGTP incorporated}}{1.2 \times 10^{12} \text{ 3' ends}} = 10$$

Keeping quantities expressed as picomoles saves a step.

Problem 6.21 For the tailing reaction experiment described in this section, how long are the dG tails?

Solution 6.21 In Problem 6.20, it was calculated that 50.4 pmol dGTP were incorporated into tails. In Problem 6.16, it was calculated that the reaction contains 2.4 pmol 3′ ends. Using these values in the equation for calculating homopolymeric tail length gives the following relationship:

$$\text{dG tail length} = \frac{50.4 \text{ pmol dGTP incorporated}}{2.4 \text{ pmol 3' ends}} = 21 \text{ residues}$$

Therefore, 21 dG residues were added onto each 3′ end.

In Vitro Transcription

In vitro RNA transcripts are synthesized for use as hybridization probes, as substrates for in vitro translation, and as antisense strands to inhibit gene expression. To monitor the efficiency of RNA synthesis, labeled CTP or UTP can be added to the reaction. Incorporation of labeled nucleotide into RNA strands is determined by the general procedure of TCA precipitation, as described earlier in this chapter. Percent incorporation can then be calculated by the following formula:

$$\% \text{ incorporation} = \frac{\text{TCA - precipitable cpm}}{\text{total cpm}} \times 100$$

Other calculations for the determination of quantity of RNA synthesized and the specific activity of the RNA product are similar to those described previously in this section and are demonstrated by the following problems.

Problem 6.22 Five hundred μCi of [α-^{35}S]UTP (1500 Ci/mmol) are added to a 20-μL in vitro transcription reaction containing 500 μM each of ATP, GTP, and CTP, 20 units of T7 RNA polymerase, and 1 μg of DNA. Following incubation of the reaction, 1 μL is diluted into a total volume of 100 μL TE buffer. Five μL of this dilution are spotted onto a filter and counted to determine total cpm. Another 5-μL sample of the dilution is precipitated with TCA and counted to determine incorporation of label. The sample prepared to determine total cpm gives 1.1×10^6 cpm. The TCA-precipitated sample gives 7.7×10^5 cpm. The RNA synthesized in this reaction is purified free of unincorporated label prior to use. **(a)** How much RNA was synthesized? **(b)** What is its specific activity?

Solution 6.22(a) The first step in solving this problem is to calculate the percent of labeled nucleotide incorporated. This is done as follows:

$$\frac{7.7 \times 10^5 \text{ TCA - precipitated cpm}}{1.1 \times 10^6 \text{ total cpm}} \times 100 = 70\% \text{ incorporation of label}$$

Therefore, 70% of the [α-^{35}S]UTP was incorporated into RNA.

The reaction contained 500 μCi of [α-^{35}S]UTP having a specific activity of 1500 Ci/mmole. This information can be used to calculate the number of nanomoles of [α-^{35}S]UTP added to the reaction as follows:

$$500 \ \mu\text{Ci } [\alpha\text{-}^{35}\text{S}]\text{UTP} \times \frac{\text{mmol}}{1500 \text{ Ci}} \times \frac{\text{Ci}}{1 \times 10^6 \ \mu\text{Ci}} \times \frac{1 \times 10^6 \text{ nmol}}{\text{mmol}} = 0.3 \text{ nmol UTP}$$

Therefore, the reaction contains 0.3 nmol UTP.

Since 70% of the UTP was incorporated into RNA, the number of nanomoles of UTP incorporated into RNA is given by the following equation:

$$0.3 \text{ nmol UTP} \times 0.7 = 0.2 \text{ nmol UTP incorporated}$$

Therefore, 0.2 nmol of UTP were incorporated into RNA.

Assuming that there are equal numbers of A's, C's, G's, and U's in the RNA transcript, then 0.2 nmol (as just calculated) represents 25% of the bases incorporated. Therefore, to calculate the total number of nanomoles of NTP incorporated, 0.2 nmol should be multiplied by 4.

$$0.2 \text{ nmol} \times 4 \text{ nucleotides} = 0.8 \text{ nmol nucleotide incorporated}$$

Using the average molecular weight of a nucleotide (330 ng/nmol), the amount of RNA synthesized can be calculated as follows:

$$0.8 \text{ nmol NTP} \times \frac{330 \text{ ng}}{\text{nmol NTP}} = 264 \text{ ng RNA sythesized}$$

Therefore, 264 ng of RNA were synthesized in the in vitro transcription reaction.

Solution 6.22(b) Specific activity is expressed in cpm/μg RNA. In part (a) of this problem, it was determined that a total of 264 ng of RNA were synthesized. To determine a specific activity, the number of cpm incorporated for the entire reaction must be calculated. A 1-μL sample of the reaction was diluted 1 μL/100 μL, and then 5 μL of this dilution were used to measure TCA-precipitable cpm. The TCA-precipitated sample gave 7.7×10^5 cpm. To calculate the incorporated cpm in the actual reaction, 7.7×10^5 cpm must be multiplied by both the dilution factor (100 μL/1 μL) and the reaction volume:

$$\frac{7.7 \times 10^5 \text{ cpm}}{5 \ \mu\text{L}} \times \frac{100 \ \mu\text{L}}{1 \ \mu\text{L}} \times 20 \ \mu\text{L} = 3.1 \times 10^8 \text{ cpm}$$

Therefore, there were a total of 3.1×10^8 cpm incorporated in the in vitro transcription reaction. Dividing this value by the total RNA synthesized and multiplying by a conversion factor to bring ng to μg gives the following equation:

$$\frac{3.1 \times 10^8 \text{ cpm}}{264 \text{ ng RNA}} \times \frac{1000 \text{ ng}}{\mu\text{g}} = \frac{1.2 \times 10^9 \text{ cpm}}{\mu\text{g RNA}}$$

Therefore, the RNA synthesized in this reaction has a specific activity of 1.2×10^9 cpm/μg RNA.

Oligonucleotide Synthesis

7

Introduction

Oligonucleotides (**oligos**), short single-stranded nucleic acids, are used for a number of applications in molecular biology and biotechnology. Oliogonculeotides are used as primers for DNA sequencing and PCR reactions, as probes for hybridization assays in gene detection, as building blocks for the construction of synthetic genes, and as antisense molecules for the control of gene expression. With such a central role in these applications as well as in many others, it is understandable that many laboratories have acquired the instrumentation and capability of synthesizing oligonucleotides for their own purposes. It is just as understandable that an entire industry has flourished to provide custom oligonucleotides to the general academic and biotechnology marketplace.

A number of methods have been described for the chemical synthesis of nucleic acids. However, almost all DNA synthesis instruments available commercially are designed to support the phosphoramidite method. In this chemistry, a phosphoramidite (a nucleoside with side protecting groups that preserve the integrity of the sugar, the phosphodiester linkage, and the base during chain extension steps) is coupled through its reactive 3′ phosphorous group to the 5′ hydroxyl goup of a nucleoside immobilized on a solid support column. The steps of oligonucleotide synthesis include the following. (1) **Detritylation**, in which the dimethoxytrityl (DMT or "trityl") group on the 5′ hydroxyl of the support nucleoside is removed by treatment with trichloroacetic acid (TCA). (2) In the **coupling step**, a phosphoramidite, made reactive by tetrazole (a weak acid), is chemically coupled to the last base added to the column support material. (3) In the **capping step**, any free 5′ hydroxyl groups of unreacted column nucleotides are acetylated by treatment with acetic anhydride and *N*-methylimidazole. (4) In the final step, called **oxidation**, the unstable internucleotide phosphite linkage between the previously coupled base and the most recently added base is oxidized by treatment with iodine and water to a more stable phosphotriester linkage. Following coupling of all bases in the oligonucleotide's sequence, the completed nucleic acid chain is cleaved from the column by treatment with ammonium hydroxide, and the base protecting groups are removed by heating in the ammonium hydroxide solution.

Synthesis Yield

Nucleic acid synthesis columns come in several scales, which are a function of the amount of the 3′ nucleoside coupled to the support material. Popular column sizes range from 40 nmol to 10 μmol. Popular support materials include controlled-pore glass (CPG) and polystyrene. The CPG columns are typically supplied having pore sizes of either 500 Å or 1000 Å. The scale of column used for a synthesis depends on both the quantity and the length of oligonucleotide desired. The greater the amount of 3′ nucleoside attached to the column support, the greater the yield of oligo. Polystyrene or large-pore CPG columns are recommended for oligonucleotides greater than 50 bases in length.

A 0.2-μmol CPG column contains 0.2 μmol of starting 3′ base. If the second base in the synthesis is coupled with 100% efficiency, then 0.2 μmol of that second base will couple, for a total of 0.4 μmol of base now on the column (0.2 μmol 3′ base + 0.2 μmol coupled base = 0.4 μmol of bases). The number of micromoles of an oligonucleotide sythesized on a column, therefore, can be calculated by multiplying the length of the oligonucleotide by the column scale.

Problem 7.1 An oligo 22 nucleotides in length (a 22-mer) is synthesized on a 0.2-μmol column. Assuming 100% synthesis efficiency, how many micromoles of oligo are made?

Solution 7.1 Multiplying the column scale by the length of the oligo yields the following equation:

$$22 \text{ base additions} \times \frac{0.2 \ \mu\text{mol}}{\text{base addition}} = 4.4 \ \mu\text{mol}$$

Therefore, assuming 100% coupling efficiency, 4.4 μmol of oligo are synthesized.

If bases are coupled at 100% efficiency, then a theoretical yield for a synthesis can be determined by using the following relationship: 1 μmol of DNA oligonucleotide contains about 10 OD units per base. (An **OD unit** is the amount of oligonucleotide dissolved in 1.0 mL, which gives an A_{260} of 1.00 in a cuvette with a 1-cm-length light path.) The theoretical yield of crude oligonucleotide is calculated by multiplying the synthesis scale by the oligo length and then multiplying the product by 10, as shown in the following problem.

Problem 7.2 A 22-mer is synthesized on a 0.2-μmol-scale column. What is the maximum theoretical yield of crude oligo in OD units?

Solution 7.2 The theoretical maximum yield of crude oligo is given by the following equation:

$$22 \text{ base additions} \times \frac{0.2 \ \mu\text{mol base}}{\text{base addition}} \times \frac{10 \ \text{OD units}}{\mu\text{mol base}} = 44 \ \text{OD units}$$

Therefore, 44 OD units of crude oligonucleotide will be synthesized.

Nucleic acid synthesis chemistry, however, is not 100% efficient. No instrument, no matter how well engineered or how pure the reagents it uses, can make oligonucleotides at the theoretical yield. The actual expected yield of synthesis product will be less. Table 7.1 lists the yield of crude oligonucleotide that can be expected from several different columns when synthesis is optimized.

Synthesis Scale	OD Units/Base
40 nmol	0.25 – 0.5
0.2 μmol	1.0 – 1.5
1 μmol	5
10 μmol	40

Table 7.1 The approximate yield, in OD units, of oligonucleotide per base for each of the common synthesis scales.

Problem 7.3 A 22-mer is synthesized on a 0.2-μmol-scale column. Assuming that synthesis is optimized, how many OD units can be expected?

Solution 7.3 Using the OD units/base value in Table 7.1 for the 0.2-μmol-scale column yields the following equations:

$$22 \ \text{bases} \times \frac{1 \ \text{OD unit}}{\text{base}} = 22 \ \text{OD units}$$

$$22 \ \text{bases} \times \frac{1.5 \ \text{OD units}}{\text{base}} = 33 \ \text{OD units}$$

Therefore, this synthesis should produce between 22 and 33 OD units of oligonucleotide.

Measuring Stepwise and Overall Yield by the DMT Cation Assay

Overall yield is the amount of full-length product synthesized. In contrast, **stepwise yield** is a measure of the percentage of the nucleic acid molecules on the synthesis column that coupled at each base addition. If the reagents and the DNA synthesizer are performing optimally, 98% stepwise yield can be expected. For such a synthesis, the 2% of molecules that do not participate in a coupling reaction will be capped and will be unavailable for reaction in any subsequent cycle. Since, at each cycle, 2% fewer molecules can react with phosphoramidite, the longer the oligo being synthesized, the lower the overall yield.

The dimethoxytrityl (DMT or trityl) group protects the 5′ hydroxyl on the ribose sugar of the most recently added base, preventing it from participating in unwanted side reactions. Prior to addition of the next base, the trityl group is removed by treatment with trichloroacetic acid (TCA). Detritylation renders the 5′ hydroxyl available for reaction with the next phosphoramidite. When removed under acidic conditions, the trityl group has a characteristic bright orange color. A fraction collector can be configured to the instrument to collect each trityl fraction as it is washed from the synthesis column during the detritylation step. The fractions can then be diluted with *p*-toluene sulfonic acid monohydrate in acetonitrile and assayed for absorbance at 498 nm. (For accuracy, further dilutions may be necessary to keep readings between 0.1 and 1.0, within the range of linear absorbance.)

For the following calculations, the first trityl fraction, that generated from the column 3′ base, should not be used, since it can give a variable reading depending on the integrity of the trityl group on the support base. Spontaneous detritylation of the column base can occur prior to synthesis, resulting in a low absorbance for the first fraction. Including this fraction in the calculation for overall yield would result in an overestimation of product yield.

Overall Yield

Overall yield, the measure of the total product synthesized, is given by the formula

$$\text{overall yield} = \frac{\text{trityl absorbance of last base added}}{\text{trityl absorbance of first base added}} \times 100$$

Problem 7.4 An 18-mer is synthesized on a 0.2-μmol column. Assay of the trityl fractions gave the following absorbance values at 498 nm.

Number	Base	Absorbance
1	Column C	0.524
2	G	0.558
3	T	0.541
4	A	0.538
5	A	0.527
6	C	0.515
7	T	0.505
8	G	0.493
9	G	0.485
10	C	0.474
11	A	0.468
12	T	0.461
13	T	0.452
14	T	0.439
15	C	0.427
16	G	0.412
17	A	0.392
18	A	0.379

What is the overall yield?

Solution 7.4 Using the equation for overall yield gives

$$\text{overall yield} = \frac{0.379}{0.558} \times 100 = 68\%$$

Therefore, 68% of the starting column base was made into full-length product; the overall yield is 68%

Stepwise Yield

Stepwise yield, the percent of molecules coupled at each single base addition, is calculated by the formula

$$\text{stepwise yield} = (\text{overall yield})^{1/\text{couplings}}$$

The number of couplings is represented by the number of trityl fractions spanned to calculate the overall yield minus 1. (One is subtracted because the trityl fraction from the first base coupled to the column base can be an unreliable indicator of actual coupling efficiency at that step.) An efficient synthesis will have a stepwise yield of 98 ± 0.5%.

A low stepwise yield can indicate reagent problems, either in their quality or in their delivery. Even if a high stepwise yield is obtained, however, the product may still not be of desired quality because some synthesis failures are not detectable by the trityl assay.

Problem 7.5 What is the stepwise yield for the synthesis described in Problem 7.4?

Solution 7.5 In Problem 7.4, an overall yield of 68% was obtained. A total of 17 trityl fractions spanned the absorbance values used to calculate the overall yield. Using the equation for stepwise yield gives

$$\text{stepwise yield} = 0.68^{1/17-1} = 0.68^{1/16} = 0.68^{0.06} = 0.98$$

($0.68^{0.06}$ is obtained on a calaculator by entering **.**, **6**, **8**, $\boldsymbol{x^y}$, **.**, **0**, **6**, and **=**.). Stepwise yield is usually expressed as a percent value:

$$0.98 \times 100 = 98\%$$

Therefore, the stepwise yield is 98%.

Assuming a stepwise yield of 98%, it is possible to calculate the overall yield for any length of oligonucleotide by using the formula

$$\text{overall yield} = 0.98^{(\text{number of bases}-1)} \times 100$$

Problem 7.6 Fresh reagents are used to synthesize a 25-mer. Assuming 98% stepwise yield, what overall yield can be expected?

Solution 7.6 The relationship just shown above yields the following equation:

$$\text{overall yield} = 0.98^{(25-1)} \times 100 = 0.98^{24} \times 100 = 0.62 \times 100 = 62\%$$

(0.98^{24} is obtained on a calculator by entering **.**, **9**, **8**, $\boldsymbol{x^y}$, **2**, **4**, and **=**.)

Therefore, an overall yield of 62% can be expected.

Problem 7.7 Assuming a stepwise yield of 95%, what overall yield can be expected for the synthesis of a 25-mer?

Solution 7.7 Using the same relationship gives

$$\text{overall yield} = 0.95^{(25-1)} \times 100 = 0.95^{24} \times 100 = 0.29 \times 100 = 29\%$$

Therefore, an overall yield of 29% can be expected for a 25-mer when the stepwise yield is 95%.

[Notice that a drop in stepwise yield from 98% (Problem 7.6) to 95% (Problem 7.7) results in a dramatic drop in overall yield from 62% to 29%. An instrument and reagents of highest quality are essential to obtaining optimal synthesis of product.]

Calculating Micromoles of Nucleoside Added at Each Base Addition Step

Trityl fraction assays can be used not only to estimate stepwise and overall yields but also to determine the number of micromoles of nucleoside added to the oligonucleotide chain at any coupling step. The number of micromoles of DMT cation released and the number of micromoles of nucleoside added are equivalent. Micromoles of DMT can be calculated by dividing a trityl fraction's absorbance at 498 by the extinction coefficient of DMT according to the formula

$$\mu\text{mol DMT} = \frac{A_{498} \times \text{volume} \times \text{dilution factor}}{e}$$

where A_{498} is the absorbance of the trityl fraction at 498 nm, volume is the final volume of the diluted DMT used for the A_{498} reading, dilution factor is the inverse of the dilution (if one is used) of the trityl fraction into a second tube of 0.1 *M* toluene sulfonic acid in acetonitrile, and $e = 70$ mL/μmol, the extinction coefficient of the DMT cation at 498 nm.

Problem 7.8 A trityl fraction from a 0.2-μmol synthesis is brought up to 10 mL with 0.1 *M* toluene sulfonic acid monohydrate in acetonitrile, and the absorbance is read at 498 nm. A reading of 0.45 is obtained. How many micromoles of nucleoside coupled at that step?

Solution 7.8 The trityl fraction was brought up to a volume of 10 mL and this volume was used to measure absorbance. No dilution into a second tube was used. Using the earlier formula yields

$$\mu\text{mol DMT} = \frac{0.45 \times 10 \text{ mL}}{70 \text{ mL}/\mu\text{mol}} = 0.06$$

Therefore, since micromoles of DMT are equivalent to micromoles of nucleoside added, 0.06 μmol of nucleoside were added at the coupling step represented by this trityl fraction.

The Polymerase Chain Reaction

8

Introduction

The **polymerase chain reaction** (**PCR**) is a method for the amplification of a specific segment of nucleic acid.* A typical PCR reaction contains a template nucleic acid (DNA or RNA), a thermally stable DNA polymerase, the four deoxynucleoside triphosphates (dNTPs), magnesium, oligonucleotide primers, and a buffer. Three steps are involved in the amplification process. (1) In a **denaturation** step, the template nucleic acid is made single-stranded by exposure to an elevated temperature (92–98°C). (2) The temperature of the reaction is then brought to between 65°C and 72°C in an **annealing** step by which the two primers attach to opposite strands of the denatured template such that their 3′ ends are directed toward each other. (3) In the final step, called **extension**, a thermally stable DNA polymerase adds dNTPs onto the 3′ ends of the two annealed primers such that strand replication occurs across the area between and including the primer annealing sites. These steps make up one **cycle** and are repeated over and over again for, depending on the specific application, from 25 to 40 times.

A very powerful technique, PCR has found use in a wide range of applications. To cite only a few, it is used to examine biological evidence in forensic cases, to identify contaminating microorganisms in food, to diagnose genetic diseases, and to map genes to specific chromosome segments.

Template and Amplification

One of the reasons the PCR technique has found such wide use is that only very little starting template DNA is required for amplification. In forensic applications, for example, a result can be obtained from the DNA recovered from a single hair or from residual saliva on a licked envelope. Typical PCR reactions performed in forensic analysis use from only 0.5 to 10 ng of DNA.

Problem 8.1 Twenty nanograms of human genomic DNA are needed for a PCR reaction. The template stock has a concentration of 0.2 mg DNA/mL. How many microliters of the stock solution should be used?

Solution 8.1 Conversion factors are used to bring mg and mL quantities to ng and μL quantities, respectively. The equation can be written as follows:

*The PCR process is covered by patents owned by Roche Molecular Systems, Inc. and F. Hoffmann-La Roche Ltd.

$$\frac{0.2 \text{ mg DNA}}{\text{mL}} \times \frac{1 \text{ mL}}{1 \times 10^3 \ \mu\text{L}} \times \frac{1 \times 10^6 \text{ ng}}{\text{mg}} \times x \ \mu\text{L} = 20 \text{ ng DNA}$$

By writing the equation in this manner, terms cancel so that the equation reduces to ng = ng. Multiplying and solving for x yields the following relationship.

$$\frac{(2 \times 10^5 \text{ ng DNA})(x)}{1 \times 10^3} = 20 \text{ ng DNA}$$

Numerator values are multiplied together, as are denominator values.

$$(2 \times 10^5 \text{ ng DNA})(x) = 2 \times 10^4 \text{ ng DNA}$$

Multiply both sides of the equation by 1×10^3.

$$x = \frac{2 \times 10^4 \text{ ng DNA}}{2 \times 10^5 \text{ ng DNA}} = 0.1$$

Divide both sides of the equation by 2×10^5 ng DNA.

Therefore, 0.1 µL of the 0.2 mg/mL human genomic DNA stock will contain 20 ng of DNA. It is often the case that a calculation is performed, such as this one, that derives an amount that cannot be measured accurately with standard laboratory equipment. Most laboratory pipettors cannot deliver 0.1 µL with acceptable accuracy. A dilution of the stock, therefore, should be performed to accommodate the accuracy limits of the available pipettors. Most laboratories have pipettors that can deliver 2 µL accurately. We can ask the question, "Into what volume can 2 µL of the stock solution be diluted such that 2 µL of the new dilution will deliver 20 ng of DNA?" The equation for this calculation is then written as follows:

$$\frac{0.2 \text{ mg DNA}}{\text{mL}} \times \frac{1 \text{ mL}}{1 \times 10^3 \ \mu\text{L}} \times \frac{1 \times 10^6 \text{ ng}}{\text{mg}} \times \frac{2 \ \mu\text{L}}{x \ \mu\text{L}} \times 2 \ \mu\text{L} = 20 \text{ ng DNA}$$

$$\frac{8 \times 10^5 \text{ ng DNA}}{(1 \times 10^3)(x)} = 20 \text{ ng DNA}$$

Cancel terms and multiply numerator values and denominator values.

$$\frac{8 \times 10^5 \text{ ng DNA}}{1 \times 10^3} = (20 \text{ ng DNA})x$$

Multiply both sides of the equation by x and solve for x.

$$800 \text{ ng DNA} = (20 \text{ ng DNA})x$$

Simplify the equation.

$$\frac{800 \text{ ng DNA}}{20 \text{ ng DNA}} = x = 40$$

Divide each side of the equation by 20 ng DNA.

Therefore, if 2 μL of the DNA stock is diluted into a total volume of 40 μL (2 μL DNA stock + 38 μL water), then 2 μL of that dilution will contain 20 ng of template DNA.

Exponential Amplification

During the PCR process, the product of one cycle serves as template in the next cycle. This characteristic is responsible for PCR's ability to amplify a single molecule into millions. PCR is an exponential process. Amplification of template progresses at a rate of 2^n, where n is equal to the number of cycles. The products of PCR are termed **amplicons**. The DNA segment amplified is called the **target**.

Problem 8.2 A single molecule of DNA is amplified by PCR for 25 cycles. Theoretically, how many molecules of amplicon will be produced?

Solution 8.2 Using the relationship of overall amplification as equal to 2^n, where n is, in this case, 25, gives the following result:

$$\text{overall amplification} = 2^{25} = 33,554,432$$

Therefore, over 33 million amplicons will be produced.

Problem 8.3 Beginning with 600 template DNA molecules, after 25 cycles of PCR, how many amplicons will be produced?

Solution 8.3 Since each template can enter into the PCR amplification process, 2^{25} should be multiplied by 600:

$$\text{overall amplification} = 600 \times 2^{25} = 2 \times 10^{10} \text{ molecules}$$

Therefore, the amplification of 600 target molecules in a PCR reaction run for 25 cycles should theoretically produce 2×10^{10} molecules of amplified product.

Problem 8.4 If the PCR reaction cited in Problem 8.2 is performed as a 100-μL reaction, how many molecules of amplified product will be present in 0.001 μL?

Solution 8.4 A ratio can be set up that is read "2×10^{10} amplicons is to 100 μL as x amplicons is to 0.001 μL."

$$\frac{2\times10^{10}\ \text{amplicons}}{100\ \mu\text{L}}=\frac{x\ \text{amplicons}}{0.001\ \mu\text{L}}$$

$$\frac{(2\times10^{10}\ \text{amplicons})(0.001\ \mu\text{L})}{100\ \mu\text{L}}=x\ \text{amplicons}$$

$$x=2\times10^{5}\ \text{amplicons}$$

A note on contamination control. As shown in Problem 8.4, very small quantities of a PCR reaction can contain large amounts of product. Laboratories engaged in the use of PCR, therefore, must take special precautions to avoid contamination of a new reaction with product from a previous one. Even quantities of PCR product present in an aerosol formed during the opening of a reaction tube are enough to overwhelm a new reaction. Contamination of a workspace by amplified product is of particular concern in facilities that amplify the same locus (or loci) over and over again, as is the case in forensic and paternity testing laboratories.

Problem 8.5 One hundred nanograms of a 242-bp fragment is produced in a PCR reaction. Two nanograms of human genomic DNA were used as starting template. What amount of amplification occurred?

Solution 8.5 The first step in solving this problem is to determine how many copies are represented by a particular amount of DNA (in this case, 2 ng of human genomic DNA). Assume that the target sequence is present in only one copy in the genome. Although there would be two copies of the target sequence in all human cells (except egg, sperm, or red blood cells), there would be one copy in the haploid human genome. The haploid human genome consists of approximately 3×10^{9} bp. Therefore, there is one copy of the target sequence per 3×10^{9} bp. The molecular weight of a base pair (660 g/mol) and Avagadro's number can be used to convert this to an amount of DNA per gene copy. Since this amount will be quite small, for convenience the answer can be expressed in picograms:

$$\frac{3\times10^{9}\ \text{bp}}{\text{copy}}\times\frac{660\ \text{g}}{\text{mol bp}}\times\frac{1\ \text{mol bp}}{6.023\times10^{23}\ \text{bp}}\times\frac{1\times10^{12}\ \text{pg}}{\text{g}}=\frac{3.3\ \text{pg}}{\text{copy}}$$

Therefore, 3.3 pg of human genomic DNA contain 1 copy of the target sequence.

The number of copies of the target sequence contained in 2 ng of human genomic DNA can now be determined:

$$2 \text{ ng} \times \frac{1 \times 10^3 \text{ pg}}{\text{ng}} \times \frac{1 \text{ copy}}{3.3 \text{ pg}} = 600 \text{ copies}$$

Therefore, 600 copies of the target sequence are contained in 2 ng of human genomic DNA template.

By a similar method, the weight of one copy of the 242-bp product can be determined as follows:

$$\frac{242 \text{ bp}}{\text{copy}} \times \frac{660 \text{ g}}{\text{mol bp}} \times \frac{1 \text{ mol bp}}{6.023 \times 10^{23} \text{ bp}} \times \frac{1 \times 10^{12} \text{ pg}}{\text{g}} = \frac{2.7 \times 10^{-7} \text{ pg}}{\text{copy}}$$

Therefore, a 242-bp fragment, representing one copy of PCR product, weighs 2.7×10^{-7} pg.

The number of copies of a 242-bp fragment represented by 100 ng of that fragment can now be determined:

$$100 \text{ ng} \times \frac{1 \times 10^3 \text{ pg}}{\text{ng}} \times \frac{\text{copy}}{2.7 \times 10^{-7} \text{ pg}} = 3.7 \times 10^{11} \text{ copies}$$

Therefore, there are 3.7×10^{11} copies of the 242-bp PCR product in 100 ng.

The amount of amplification can now be calculated by dividing the number of copies of the target sequence at the end of the PCR reaction by the number of copies of target present in the input template DNA:

$$\frac{3.7 \times 10^{11} \text{ copies}}{600 \text{ copies}} = 6.2 \times 10^8$$

Therefore, the PCR reaction created a (6.2×10^8)-fold increase in the amount of amplified DNA segment.

PCR Efficiency

The formula 2^n used to calculate PCR amplification can be employed to describe any system that proceeds by a series of doubling events, whereby 1 gives rise to 2, 2 gives rise to 4, 4 to 8, 8 to 16, 16 to 32, etc. This relationship describes the PCR if the reaction is 100% efficient. No reaction performed in a test tube, however, will be 100% efficient. To more accurately represent reality, the relationship must be modified to account for actual reaction efficiency. The equation then becomes

$$Y = (1 + E)^n$$

where

Y = degree of amplification
n = number of cycles
E = mean efficiency of amplification at each cycle; the percent of product from one cycle serving as template in the next

Loss of amplification efficiency can occur for any number of reasons, including loss of DNA polymerase enzyme activity, depletion of reagents, and incomplete denaturation of template, as occurs when the amount of product increases in later cycles. Efficiency of amplification can be experimentally determined by taking a small aliquot out of the PCR reaction at the beginning and end of each cycle (from the third cycle on) and electrophoresing that sample on a gel. In other lanes on that gel are run standards with known numbers of molecules. Intensities of the bands (as determined by staining or densitometry) are compared for quantitation. The efficiency, expressed as a decimal, is then the final copy number at the end of a cycle divided by the number of copies at the beginning of that cycle.

To take into account the amount of starting template in a PCR (represented by X), the equation becomes

$$Y = X(1+E)^n$$

Use of this relationship requires an understanding of logarithms and antilogarithms. A **common logarithm**, as will be used here, is defined as the exponent indicating the power to which 10 must be raised to produce a given number. An **antilogarithm** is found by reversing the process used to find the logarithm of a number; e.g., if $y = \log x$, then $x = \text{antilog } y$. Before this equation can be converted to a more useful form, two laws of logarithms must be stated, the **product rule for logarithms** and the **power rule for logarithms**. (More discussions on logarithms can be found in Chapter 3.)

Product Rule for Logarithms

For any positive numbers M, N, and a (where a is not equal to 1), the logarithm of a product is the sum of the logarithms of the factors.

$$\log_a MN = \log_a M + \log_a N$$

Power Rule for Logarithms

For any positive number M and a (where a is not equal to 1), the logarithm of a power of M is the exponent times the logarithm of M.

$$\log_a M^p = p \log_a M$$

The equation $Y = X(1 + E)^n$ can now be written in a form that will allow more direct calculation of matters important in PCR. Taking the common logarithm of both sides of the equation and using both the product and power rules for logarithms yields

$$\log Y = \log X + n \log(1 + E)$$

This is the form of the equation that can be used when calculating product yield or when calculating the number of cycles it takes to generate a certain amount of product from an initial template concentration.

Problem 8.6 An aliquot of template DNA containing 3×10^5 copies of a target gene is placed into a PCR reaction. The reaction has a mean efficiency of 85%. How many cycles are required to produce 2×10^{10} copies?

Solution 8.6 For the equation $Y = X(1 + E)^n$, $Y = 2 \times 10^{10}$ copies, $X = 3 \times 10^5$ copies, and $E = 0.85$. Placing these values into the equation gives the following relationship.

$$2 \times 10^{10} = (3 \times 10^5)(1 + 0.85)^n$$

$$\frac{2 \times 10^{10}}{3 \times 10^5} = 1.85^n$$

Divide each side of the equation by 3×10^5.

$$6.7 \times 10^4 = 1.85^n$$

Simplify the left side of the equation.

$$\log 6.7 \times 10^4 = \log 1.85^n$$

Take the common logarithm of both sides of the equation.

$$\log 6.7 \times 10^4 = n \log 1.85$$

Use the power rule for logarithms.

$$\frac{\log 6.7 \times 10^4}{\log 1.85} = n$$

Divide both sides of the equation by log 1.85.

$$\frac{4.8}{0.27} = n = 17.8$$

Substitute the log values into the denominator and numerator. (To find the log of a number on a calculator, enter that number and then press the ***log*** key.)

Therefore, it will take 18 cycles to produce 2×10^{10} copies of amplicon starting from 3×10^5 copies of template.

Problem 8.7 A PCR amplification proceeds for 25 cycles. The initial target sequence was present in the reaction at 3×10^5 copies. At the end of the 25 cycles, 4×10^{12} copies were produced. What was the efficiency (E) of the reaction?

Solution 8.7 Using the equation $Y = X(1 + E)^n$, $Y = 4\times10^{12}$ copies, $X = 3\times10^5$ copies, and $n = 25$, the equation then becomes

$$4\times10^{12} = (3\times10^5)(1+E)^{25}$$

$$\frac{4\times10^{12}}{3\times10^5} = (1+E)^{25}$$

Divide each side of the equation by 3×10^5 and simplify.

$$1.3\times10^7 = (1+E)^{25}$$

$$\log 1.3\times10^7 = \log(1+E)^{25}$$

Take the logarithm of both sides of the equation.

$$\log 1.3\times10^7 = 25 \log(1+E)$$

Use the power rule for logarithms.

$$7.1 = 25 \log(1+E)$$

Substitute the log of 1.3×10^7 into the equation (The log of 1.3×10^7 is found on a calculator by entering **1**, **.**, **3**, **EXP**, **7**, **log**.)

$$\frac{7.1}{25} = \log(1+E)$$

Divide each side of the equation by 25 and simplify the left side of the equation.

$$0.28 = \log(1+E)$$

$$\text{antilog } 0.28 = \text{antilog}[\log(1+E)]$$

Take the antilog of each side of the equation.

$$\text{antilog } 0.28 = 1+E$$

$$1.91 = 1+E$$

To find the antilog of 0.28 on the calculator, enter **10**, $\boldsymbol{x^y}$, **.28**, and **=**.

$$1.91 - 1 = E$$

$$0.91 = E$$

Subtract 1 from both sides of the equation.

Therefore, the PCR reaction was 91% efficient ($0.91 \times 100 = 91\%$).

Calculating the T_m of the Target Sequence

The T_m (**melting temperature**) of double-stranded DNA is that temperature at which half of the molecule is in double-stranded conformation and half is in single-stranded form. It doesn't matter where along the length of the molecule the single-stranded or double-stranded regions occur so long as half of the entire molecule is in single-stranded form. It is desirable to know the melting temperature of the amplicon to help ensure that an adequate denaturation temperature is chosen for thermal cycling, one that will provide an optimal amplification reaction.

DNA is called a **polyanion** because of the multiple negative charges along the sugar–phosphate backbone. The two strands of DNA, since they carry like charges, have a natural tendency to repel each other. In a water environment free of any ions, DNA will denature easily. Salt ions such as sodium (Na^+) and magnesium (Mg^{+2}), however, can complex to the negative charges on the phosphate groups of DNA and will counteract the tendency of the two strands to repel each other. In fact, under conditions of high salt, two strands of DNA can be made to anneal to each other even if they contain a number of mismatched bases between them.

Most PCR reactions contain salt. Potassium chloride (KCl) is frequently added to a PCR reaction to enhance the activity of the polymerase. Magnesium is a cofactor of the DNA polymerase enzyme. It is provided in a PCR reaction by the salt magnesium chloride ($MgCl_2$). The presence of potassium (K^+) and magnesium ions will affect the melting temperature of duplex nucleic acid and should be a consideration when making a best calculation of T_m.

Other factors affect the T_m of duplex DNA. Longer molecules, having more base pairs holding them together, require more energy to denature them. Likewise, T_m is affected by the G + C content of the DNA. Since G's and C's form three hydrogen bonds between them (as compared with two hydrogen bonds formed between A's and T's), DNAs having higher G + C content will require more energy (a higher temperature) to denature them.

These considerations are taken into account in the formula given by Wetmur and Sninsky (1995) for calculating the melting temperature of long duplex DNAs:

$$T_m = 81.5 + 16.6 \ \log\left[\frac{[\mathrm{SALT}]}{1.0 + 0.7[\mathrm{SALT}]}\right] + 0.41(\%\mathrm{G} + \mathrm{C}) - \frac{500}{L}$$

where L = the length of the amplicon in bp, %G + C is calculated using the formula

$$\%\mathrm{G} + \mathrm{C} = \frac{\text{G + C bases in the amplicon}}{\text{total number of bases in the amplicon}} \times 100$$

and

$$[\text{SALT}] = [\text{K}^+] + 4[\text{Mg}^{2+}]^{0.5}$$

In this last equation, any other monovalent cation, such as Na^+, if in the PCR reaction, can be substituted for K^+.

For the thermal cycling program, a denaturation temperature should be chosen 3–4°C higher than the calculated T_m to ensure adequate separation of the two amplicon strands.

Problem 8.8 An 800-bp PCR product having a G + C content of 55% is synthesized in a reaction containing 50 m*M* KCl and 2.5 m*M* $MgCl_2$. What is the amplicon's T_m?

Solution 8.8 First, calculate the salt concentration ([SALT]). The formula given for doing this requires that the concentrations of K^+ and Mg^{+2} be expressed in *M* rather than m*M* concentration. For KCl this is calculated as follows:

$$50 \text{ m}M \text{ KCl} \times \frac{1 \ M}{1000 \text{ m}M} = 0.05 \ M \text{ KCl}$$

Therefore, the KCl concentration is 0.05 *M*.

The *M* concentration of $MgCl_2$ is calculated as follows:

$$2.5 \text{ m}M \text{ MgCl}_2 \times \frac{1 \ M}{1000 \text{ m}M} = 0.0025 \ M \text{ MgCl}_2$$

Therefore, the $MgCl_2$ concentration is 0.0025 *M*.

Placing these values into the equation for calculating salt concentration gives the following relationship:

$$[\text{SALT}] = [0.05] + 4[0.0025]^{0.5}$$

$[0.0025]^{0.5}$ on the calculator is obtained by entering **0.0025**, pressing the x^y key, entering **0.5**, and pressing the = key.

$$[\text{SALT}] = 0.05 + 4(0.05)$$

Solving this equation gives the following result:

$$[\text{SALT}] = 0.05 + 0.2 = 0.25$$

Placing this value into the equation for T_m yields the following equation:

$$T_m = 81.5 + 16.6 \ \log\left[\frac{[0.25]}{1.0 + 0.7[0.25]}\right] + 0.41(55) - \frac{500}{800}$$

The equation is simplified to give the following relationship:

$$T_m = 81.5 + 16.6 \ \log[0.213] + 0.41(55) - 0.63$$

Calculating the log of 0.213 (enter **0.213** on the calculator and press the ***log*** key) and further simplifying the equation gives the following result:

$$T_m = 81.5 + 16.6(-0.672) + 22.55 - 0.63$$

$$T_m = 81.5 + (-11.155) + 22.55 - 0.63 = 92.3$$

Therefore, the melting temperature of the 800-bp amplicon is 92.3°C. Programming the thermal cycler for a denaturation temperature of 96°C should ensure proper denaturation at each cycle.

Primers

PCR primers are used in concentrations ranging from 0.05 μM to 2 μM.

Problem 8.9 A primer stock consisting of the two primers needed for a PCR reaction has a concentration of 25 *μM*. What volume of the primer stock should be added to a 100-μL reaction so that the primer concentration in the reaction is 0.4 μM?

Solution 8.9 The calculation is set up in the following manner:

$$25 \ \mu M \ \text{primer} \times \frac{x \ \mu\text{L}}{100 \ \mu\text{L}} = 0.4 \ \mu M \ \text{primer}$$

The equation is then solved for x.

$$\frac{(25 \ \mu M)x}{100} = 0.4 \ \mu M$$

$$(25 \ \mu M)x = 40 \ \mu M$$

$$x = \frac{40 \ \mu M}{25 \ \mu M} = 1.6$$

Therefore, 1.6 μL of the 25 μ*M* primer stock should be added to a 100-μL PCR reaction to give a primer concentration of 0.4 μ*M*.

Problem 8.10 A 100-μL PCR reaction contains primers at a concentration of 0.4 μ*M*. How many picomoles of primer are in the reaction?

Solution 8.10 Remember, 0.4 μ*M* is equivalent to 0.4 μmol/L. A series of conversion factors is used to bring a molar concentration into a mole quantity:

$$\frac{0.4\ \mu\text{mol primer}}{\text{L}} \times \frac{1\ \text{L}}{1\times 10^6\ \mu\text{L}} \times \frac{1\times 10^6\ \text{pmol}}{\mu\text{mol}} \times 100\ \mu\text{L} = 40\ \text{pmol primer}$$

Therefore, there are 40 pmol of primer in the reaction.

Problem 8.11 A 100-μL PCR reaction is designed to produce a 500-bp fragment. The reaction contains 0.2 μ*M* primer. Assuming a 100% efficient reaction and that the other reaction components are not limiting, what is the theoretical maximum amount of product (in μg) that can be made?

Solution 8.11 The first step is to calculate how many micromoles of primer are in the reaction. (0.2 μ*M* is equivalent to 0.2 μmol/L.)

$$\frac{0.2\ \mu\text{mol primer}}{\text{L}} \times \frac{1\ \text{L}}{1\times 10^6\ \mu\text{L}} \times 100\ \mu\text{L} = 2\times 10^{-5}\ \mu\text{mol primer}$$

Therefore, the reaction contains 2×10^{-5} μmol of primer. Assuming a 100% efficient reaction, each primer will be extended into a complete DNA strand. Therefore, 2×10^{-5} μmol of primer will make 2×10^{-5} μmol of extended products. Since in this case the product is 500 bp in length, its molecular weight must be part of the equation used to calculate the amount of total product made:

$$500\ \text{bp} \times \frac{660\ \text{g/mol}}{\text{bp}} \times \frac{1\ \text{mol}}{1\times 10^6\ \mu\text{mol}} \times \frac{1\times 10^6\ \mu\text{g}}{\text{g}} \times 2\times 10^{-5}\ \mu\text{mol} = 6.6\ \mu\text{g}$$

Therefore, in this reaction, 6.6 μg of 500-bp PCR product can be synthesized.

Problem 8.12 Twenty pg of an 8000-bp recombinant plasmid are used in a 50-μL PCR reaction in which, after 25 cycles, 1 μg of a 400-bp PCR fragment is produced. The reaction contained 0.2 μ*M* primer. (**a**) How much of the primer was consumed during the synthesis of the product? (**b**) How much primer remains after 25 cycles? (**c**) How many cycles would be required to completely consume the entire reaction's supply of primer?

Solution 8.12(a) Since each PCR fragment contains primer incorporated into each strand in a ratio of 1 primer to 1 DNA strand, the picomolar concentration of PCR fragment will be equivalent to the picomolar concentration of primer. The first step is to calculate the molecular weight of the 400-bp PCR fragment. This is accomplished by multiplying the size of the fragment by the molecular weight of a single base pair:

$$400 \text{ bp} \times \frac{660 \text{ g/mol}}{\text{bp}} = \frac{2.64 \times 10^5 \text{ g}}{\text{mol}}$$

Therefore, the 400-bp PCR fragment has a molecular weight of 2.64×10^5 g/mol. This value can then be used in an equation to convert the amount of 400-bp fragment produced into picomoles:

$$1 \text{ μg} \times \frac{1 \text{ g}}{1 \times 10^6 \text{ μg}} \times \frac{1 \text{ mol}}{2.64 \times 10^5 \text{ g}} \times \frac{1 \times 10^{12} \text{ pmol}}{\text{mol}} = 3.8 \text{ pmol}$$

Therefore, 3.8 pmol of 400-bp fragment were synthesized in the PCR reaction. If 3.8 pmol of fragment were produced, then 3.8 pmol of primer were consumed by incorporation into product.

Solution 8.12(b) To calculate how much primer remains after 25 cycles, we must first calculate how much primer we started with. Primer was in the reaction at a concentration of 0.2 μ*M*. We can calculate how many picomoles of primer are represented by 0.2 μ*M* primer in the following manner: (Remember, 0.2 μ*M* primer is equivalent to 0.2 μmol primer/L.)

$$\frac{0.2 \text{ μmol primer}}{\text{L}} \times \frac{1 \times 10^6 \text{ pmol}}{\text{μmol}} \times \frac{1 \text{ L}}{1 \times 10^6 \text{ μL}} \times 50 \text{ μL} = 10 \text{ pmol primer}$$

Therefore, the reaction began with 10 pmol of primer. And 3.8 pmol of primer were consumed in the production of PCR fragment (Solution 8.12a). This amount is subtracted from the 10 pmol of primer initially contained in the reaction:

$$10 \text{ pmol} - 3.8 \text{ pmol} = 6.2 \text{ pmol}$$

Therefore, 6.2 pmol of primer remain unincorporated after 25 cycles.

Solution 8.12(c) The solution to this problem requires the use of the general equation, $Y = X(1 + E)^n$. We will solve for n, the exponent. As for the other components, from the information given we need to determine Y, the ending number of copies of fragment, X, the initial number of target molecules, and E, the overall efficiency of the reaction. The starting number of target sequence is calculated by first determining the weight of one copy of the 8000-bp recombinant plasmid:

$$\frac{8000 \text{ bp}}{\text{copy}} \times \frac{660 \text{ g}}{\text{mol bp}} \times \frac{1 \text{ mol bp}}{6.023 \times 10^{23} \text{ bp}} = \frac{8.8 \times 10^{-18} \text{ g}}{\text{copy}}$$

Therefore, each copy of the 8000-bp recombinant plasmid weighs 8.8×10^{-18} g. This value can then be used as a conversion factor to determine how many copies of the recombinant plasmid are represented by 20 pg, the starting amount of template:

$$20 \text{ pg} \times \frac{\text{g}}{1 \times 10^{12} \text{ pg}} \times \frac{\text{copy}}{8.8 \times 10^{-18} \text{ g}} = 2.3 \times 10^{6} \text{ copies}$$

Therefore, there are 2.3×10^6 copies of starting template.

After 25 cycles, 1 μg of 400-bp product was made. To determine how many copies this represents, it must first be determined how much one copy weighs:

$$\frac{400 \text{ bp}}{\text{copy}} \times \frac{660 \text{ g}}{\text{mol bp}} \times \frac{1 \text{ mol bp}}{6.023 \times 10^{23} \text{ bp}} = \frac{4.4 \times 10^{-19} \text{ g}}{\text{copy}}$$

Therefore, each 400-bp PCR fragment weighs 4.4×10^{-19} g. Using this value as a conversion factor, 1 μg of product can be converted to number of copies in the following way:

$$1 \text{ μg} \times \frac{\text{g}}{1 \times 10^{6} \text{ μg}} \times \frac{\text{copy}}{4.4 \times 10^{-19} \text{ g}} = 2.3 \times 10^{12} \text{ copies}$$

Therefore, after 25 cycles, 2.3×10^{12} copies of the 400-bp PCR fragment have been made.

Using the starting number of copies of template and the number of copies of PCR fragment made after 25 cycles, the efficiency (E) of the reaction can now be calculated.

$$2.3 \times 10^{12} \text{ copies} = 2.3 \times 10^{6} \text{ copies}(1 + E)^{25}$$

$$\frac{2.3 \times 10^{12}}{2.3 \times 10^{6}} = (1 + E)^{25}$$

Divide each side of the equation by 2.3×10^6.

$$1 \times 10^{6} = (1 + E)^{25}$$

$$\log 1 \times 10^{6} = 25 \log(1 + E)$$

Take the logarithm of both sides of the equation and use the power rule for logarithms.

$$\frac{6}{25} = \log(1 + E)$$

Take the log of 1×10^6 (on the calculator, enter **1**, press the **EXP** key, enter **6**, then press the ***log*** key) and divide each side of the equation by 25.

$$0.24 = \log(1 + E)$$

$$1.74 = 1 + E$$

Take the antilog of of both sides of the equation. To find the antilog of 0.24 on the calculator, enter **0.24**, then the **10^x^** key.

$$1.74 - 1 = E$$

Subtract 1 from both sides of the equation to find *E*.

$$0.74 = E$$

Therefore, *E* has a value of 0.74.

The reaction contains 10 pmol of primer. Ten picomoles of primer will make 10 pmol of product. The number of copies this represents is calculated as follows:

$$10 \text{ pmol} \times \frac{1 \text{ mol}}{1 \times 10^{12} \text{ pmol}} \times \frac{6.023 \times 10^{23} \text{ copies}}{\text{mol}} = 6.023 \times 10^{12} \text{ copies}$$

Therefore, 10 pmol of primer can make 6.023×10^{12} copies of PCR fragment.

We now have all the information we need to calculate the number of cycles required to make 6.023×10^{12} copies (10 pmol) of product. For the following equation, *Y*, the final number of copies of product, is 6.023×10^{12} copies, *X*, the initial number of copies of target sequence, is 2.3×10^6 copies, and *E*, the efficiency of the PCR reaction, is 0.74. We will solve for the exponent *n*, the number of PCR cycles, as follows:

$$6.023 \times 10^{12} = (2.3 \times 10^6)(1 + 0.74)^n$$

$$\frac{6.023 \times 10^{12}}{2.3 \times 10^6} = (1.74)^n$$

Divide each side of the equation by 2.3×10^6 and simplify the term in parentheses.

$$2.62 \times 10^6 = (1.74)^n$$

$$\log\ 2.62 \times 10^6 = n\ \log\ 1.74$$

Take the log of both sides of the equation and use the power rule for logarithms.

$$\frac{\log\ 2.62 \times 10^6}{\log\ 1.74} = n$$

Divide each side of the equation by log 1.74.

$$\frac{6.42}{0.24} = n = 26.75$$

Take the log of the values on the left side of the equation and divide these values to give *n*.

Therefore, after 27 cycles, the 10 pmol of primer will have been converted into 10 pmol of PCR fragment. Many protocols use primer quantities as small as 10 pmol. Using a relatively small amount of primer can increase a reaction's specificity. For the reaction in this example, using more than 27 cycles would provide no added benefit.

Primer T_m

For an optimal PCR reaction, the two primers used in a PCR reaction should have similar melting temperatures. The optimal annealing temperature for a PCR experiment may actually be several degrees above or below that of the T_m of the least stable primer. The best annealing temperature should be derived empirically. The highest annealing temperature giving the highest yield of desired product and the least amount of background amplification should be used for routine amplification of a particular target. The formulas described next can be used to give the experimenter a good idea of where to start.

A number of different methods for calculating primer T_m can be found in the literature, and several are commonly used. A very quick method for estimating the T_m of an oligonucleotide primer is to add 2°C for each A or T nucleotide present in the primer and 4°C for each G or C nucleotide. Expressed as an equation, this relationship is

$$T_m = 2°(\mathrm{A} + \mathrm{T}) + 4°(\mathrm{G} + \mathrm{C})$$

This method of calculating T_m, however, is valid only for primers between 14 and 20 nucleotides in length.

Problem 8.13 A primer with the sequence ATCGGTAACGATTACATTC is to be used in a PCR reaction. Using the formula just presented, what is the T_m of this oligonucleotide?

Solution 8.13. The oligo contains 12 A's and T's and 7 G's and C's.

$$2° \times 12 = 24° \text{ and } 4° \times 7 = 28°$$

Adding these two values together gives

$$24° + 28° = 52°$$

Therefore, by this method of calculating melting temperature, the primer has a T_m of 52°C.

Calculating T_m Based on Salt Concentration, G/C Content, and DNA Length

Another formula frequently used to calculate T_m is the following:

$$T_m = 81.5° + 16.6(\log[M^+]) + 0.41(\%G + C) - (500/N)$$

where $[M^+]$ is the concentration of monovalent cation in the PCR reaction (for most PCR reactions, this would be KCl) expressed in moles per liter and N is equal to the number of nucleotides in the primer. The %G + C term in this equation should *not* be expressed as a decimal (50% G + C, not 0.50 G + C). This formula takes into account the salt concentration, the length of the primer, and the G + C content. It can be used for primers that are 14–70 nucleotides in length.

Problem 8.14 A primer with the sequence ATCGGTAACGATTACATTC is to be used in a PCR reaction. The PCR reaction contains 0.05 *M* KCl. Using the formula just given, what is the T_m of this oligonucleotide?

Solution 8.14 A 0.05 *M* KCl solution is equivalent to 0.05 mol KCl/L. The primer, a 19-mer, has 7 Gs and Cs. It has a %G + C content of 37%.

$$\frac{7 \text{ nucleotides}}{19 \text{ nucleotides}} \times 100 = 37\%$$

The T_m of the primer is calculated as follows.

$$T_m = 81.5° + 16.6(\log 0.05) + 0.41(37) - 500/19$$

$$T_m = 81.5° + 16.6(-1.3) + 15.2 - 26.3$$

$$T_m = 81.5° + (-21.6) + 15.2 - 26.3 = 48.8°\text{C}$$

Therefore, by this method of calculation, the primer has a T_m of 48.8°C. (Compare this with 52°C calculated in Problem 8.13. for the same primer sequence.)

Calculating T_m Based on Nearest-Neighbor Interactions

Wetmur and Sninsky (1995) describe a formula for calculating primer T_m that uses nearest-neighbor interactions between adjacent nucleotides within the oligonucleotide. Experiments by Breslauer et al. (1986) demonstrated that it is base sequence rather than base composition that dictates the stability and melting characteristics of a particular DNA molecule. For example, a DNA molecule having the structure

$$\begin{matrix} {}^{5'}AT^{3'} \\ {}_{3'}TA_{5'} \end{matrix}$$

will have different melting and stability properties than a DNA molecule with the structure

$$\begin{matrix} {}^{5'}TA^{3'} \\ {}_{3'}AT_{5'} \end{matrix}$$

even though they have the same base composition. The stability of a DNA molecule is determined by its free energy ($\Delta G°$). The melting behavior of a DNA molecule is dictated by its enthalpy ($\Delta H°$).

Free energy is a measure of the tendency of a chemical system to react or change. It is expressed in kilocalories per mole (kcal/mol). (A **calorie** is the amount of heat at a pressure of one atmosphere required to raise the temperature of one gram of water by one degree C.) In a chemical reaction, the energy released or absorbed in that reaction is the difference (Δ) between the energy of the reaction products and the energy of the reactants. Under conditions of constant temperature and pressure, the energy difference is defined as **ΔG** (the **Gibbs free-energy change**). The **standard free-energy change, $\Delta G°$**, is the change in free energy when one mole of a compound is formed at 25°C at one atmosphere pressure.

Enthalpy is a measure of the heat content of a chemical or physical system. The **enthalpy change, $\Delta H°$**, is the quantity of heat released or absorbed at constant temperature, pressure, and volume. It is equal to the enthalpy of the reaction products minus the enthalpy of the reactants. It is expressed in units of kcal/mol.

Tables 8.1 and 8.2, of nearest-neighbor interactions for enthalpy and free energy, are taken from Quartin and Wetmur (1989).

First Nucleotide	Second Nucleotide			
	dA	dC	dG	dT
dA	–9.1	–6.5	–7.8	–8.6
dC	–5.8	–11.0	–11.9	–7.8
dG	–5.6	–11.1	–11.0	–6.5
dT	–6.0	–5.6	–5.8	–9.1

Table 8.1 Enthalpy ($\Delta H°$) values for nearest-neighbor nucleotides (in kcal/mol).

First Nucleotide	Second Nucleotide			
	dA	dC	dG	dT
dA	–1.55	–1.40	–1.45	–1.25
dC	–1.15	–2.30	–3.05	–1.45
dG	–1.15	–2.70	–2.30	–1.40
dT	–0.85	–1.15	–1.15	–1.55

Table 8.2 Free-energy ($\Delta G°$) values for nearest-neighbor nucleotides (in kcal/mol).

Free energy and enthalpy of nearest-neighbor interactions are incorporated into the following equation for calculating primer melting temperature (Wetmur and Sninsky, 1995):

$$T_m = \frac{T° \ \Delta H°}{\Delta H° - \Delta G° + RT° \ \ln(C)} + 16.6 \ \log\left[\frac{(\text{SALT})}{1.0 + 0.7(\text{SALT})}\right] - 269.3$$

where

$$(\text{SALT}) = (\text{K}^+) + 4(\text{Mg}^{+2})^{0.5}$$

$T°$ is the temperature in degrees Kelvin (K). Standard state (25°C) is assumed for this calculation. Actual degree increments are equivalent between K and °C. 0°C is equivalent to 273.2 K. $T°$ in K, therefore, is 25° + 273.2° = 298.2 K.) R, the molar gas constant, is 1.99 cal/mol K. C is the molar concentration of primer. $\Delta H°$ is calculated as the sum of all nearest-neighbor $\Delta H°$ values for the primer plus the enthalpy contributed by "dangling end" interactions (template sequence extending past the bases complementary to the 5′ and 3′ ends of the primer). A primer will have two dangling ends when annealing to the template. The enthalpy contribution ($\Delta H°_e$) from a single dangling end is –5000 cal/mol. $\Delta H°$ is calculated by the following formula:

$$\Delta H° = \Sigma_{nn}(\text{each} \ \ \Delta H°_{nn}) + \Delta H°_e$$

(The subscript "*nn*" designates-nearest neighbor interaction. The subscript "*e*" designates the dangling-end interaction.)

$\Delta G°$, the free energy, is calculated as the sum of all nearest-neighbor $\Delta G°$ values for the primer plus free energy contributed by the initiation of base-pair formation (+2200 cal/mol) plus the $\Delta G°$ contributed by a single dangling end (–1000 cal/mol). It is calculated by the following formula:

$$\Delta G° = \Sigma_{nn}(\text{each } \Delta G°_{nn}) + \Delta G°_{e} + \Delta G°_{i}$$

(The subscript "*nn*" designates nearest-neighbor interaction. The subscripts "*e*" and "*i*" designate dangling-end interactions and initiation of base-pair formation, respectively.)

Problem 8.15 A primer with the sequence ATCGGTAACGATTACATTC is to be used in a PCR reaction at a concentration of 0.4 μ*M*. The PCR reaction contains 0.05 *M* KCl and 2.5 m*M* Mg^{2+}. Assume two dangling ends. Using the formula given earlier, what is the oligonucleotide's T_m?

Solution 8.15 The salt concentrations (K^+ and Mg^{2+}) must both be expressed as molar concentrations. 2.5 m*M* Mg^{2+} is equivalent to 0.0025 *M*, as shown here:

$$2.5 \text{ m}M \text{ Mg}^{2+} \times \frac{1 \text{ } M}{1000 \text{ m}M} = 0.0025 \text{ } M \text{ Mg}^{2+}$$

The total salt concentration is therefore

$$(\text{SALT}) = (0.05) + 4(0.0025)^{0.5} = (0.05) + (0.2) = 0.25$$

As shown in the following table, the sum of the $\Delta H°$ nearest-neighbor interactions is –142 kcal/mol.

Nearest-Neighbor Bases	$\Delta H°$
AT	–8.6
TC	–5.6
CG	–11.9
GG	–11.0
GT	–6.5
TA	–6.0
AA	–9.1
AC	–6.5
CG	–11.9
GA	–5.6
AT	–8.6
TT	–9.1
TA	–6.0
AC	–6.5
CA	–5.8
AT	–8.6
TT	–9.1
TC	–5.6
Total $\Delta H°$	**–142.0**

The $\Delta H°_e$ value is calculated as follows:

$$\frac{-5000 \text{ cal/mol}}{\text{dangling end}} \times 2 \text{ dangling ends} = -10,000 \text{ cal/mol}$$

This is equivalent to –10 kcal/mol:

$$-10,000 \text{ cal/mol} \times \frac{\text{kcal/mol}}{1000 \text{ cal/mol}} = -10 \text{ kcal/mol}$$

The total $\Delta H°$ is calculated as follows:

$$\Delta H° = -142 \text{ kcal/mol} + (-10 \text{ kcal/mol}) = -152 \text{ kcal/mol}$$

Therefore, the $\Delta H°$ for this primer is –152 kcal/mol. This is equivalent to –152,000 cal/mol:

$$-152 \text{ kcal/mol} \times \frac{1000 \text{ cal/mol}}{\text{kcal/mol}} = -152,000 \text{ cal/mol}$$

As shown in the following table, the sum of the $\Delta G°$ nearest-neighbor interactions is –27.3 kcal/mol.

Nearest-Neighbor Bases	**$\Delta G°_{nn}$**
AT	–1.25
TC	–1.15
CG	–3.05
GG	–2.30
GT	–1.40
TA	–0.85
AA	–1.55
AC	–1.40
CG	–3.05
GA	–1.15
AT	–1.25
TT	–1.55
TA	–0.85
AC	–1.40
CA	–1.15
AT	–1.25
TT	–1.55
TC	–1.15
Total ΔG°	**–27.3**

The $\Delta G°$ for the contribution from the dangling ends is determined as follows:

$$\Delta G°_e = 2 \ \text{dangling ends} \times \frac{-1000 \ \text{cal/mol}}{\text{dangling end}} \times \frac{\text{kcal/mol}}{1000 \ \text{cal/mol}} = -2 \ \text{kcal/mol}$$

$\Delta G°$ is then calculated by adding the values determined earlier to the $\Delta G°$ for initiation of base-pair formation ($\Delta G°_i$ = +2200 cal/mol = 2.2 kcal/mol).

$$\Delta G° = -27.3 \ \text{kcal/mol} + (-2 \ \text{kcal/mol}) + 2.2 \ \text{kcal/mol} = -27.1 \ \text{kcal/mol}$$

Therefore, $\Delta G°$ for this primer is –27.1 kcal/mol, which is equivalent to –27,100 cal/mol:

$$-27.1 \ \text{kcal/mol} \times \frac{1000 \ \text{cal/mol}}{\text{kcal/mol}} = -27{,}100 \ \text{cal/mol}$$

The primer is at a concentration of 0.4 μM. This is converted to a molar (M) concentration as follows:

$$0.4\ \mu M \times \frac{1\ M}{1 \times 10^{6}\ \mu M} = 4 \times 10^{-7}\ M$$

Placing these values into the Wetmur and Sninsky equation gives the following relationship:

$$T_m = \frac{T^\circ\ \Delta H^\circ}{\Delta H^\circ - \Delta G^\circ + RT^\circ\ \ln(C)} + 16.6\ \log\left[\frac{(\text{SALT})}{1.0 + 0.7(\text{SALT})}\right] - 269.3$$

$$T_m = \frac{298.2 \times (-152{,}000)}{-152{,}000 - (-27{,}100) + 1.99(298.2)\ln(4 \times 10^{-7})} + 16.6\ \log\left[\frac{(0.25)}{1.0 + 0.7(0.25)}\right] - 269.3$$

$$T_m = \frac{-45{,}326{,}400}{-152{,}000 + 27{,}100 + (593.4)\ln(4 \times 10^{-7})} + 16.6\ \log[0.21] - 269.3$$

$$T_m = \frac{-45{,}326{,}400}{-124{,}900 + (593.4)(-14.73)} + (16.6)(-0.68) - 269.3$$

$$T_m = \frac{-45{,}326{,}400}{-124{,}900 + (-8{,}741)} + (-11.3) - 269.3$$

$$T_m = 339.2 - 11.3 - 269.3 = 58.6^\circ$$

Therefore, by this method of calculation, the primer has a T_m of 58.6°C.

dNTPs

Deoxynucleoside triphosphates (dATP, dCTP, dGTP, and dTTP) are the building blocks of DNA. They should be added to a PCR reaction in equimolar amounts and, depending on the specific application, are used in concentrations ranging from 20 to 200 μM each. In preparing a series of PCR reactions, the four dNTPs can be prepared as a master mix, an aliquot of which is added to each PCR reaction in the series.

Problem 8.16 A dNTP master mix is prepared by combining 40 μL of each 10 m*M* dNTP stock. Four μL of this dNTP master mix are added to a PCR reaction having a final volume of 50 μL. (**a**) What is the concentration of each dNTP in the master mix? (**b**) What is the concentration of total dNTP in the PCR reaction? Express the concentration in micromolarity.

Solution 8.16(a) The dNTP master mix contains 40 μL of each of the four dNTPs. The total volume of the dNTP master mix, therefore, is 4 × 40 μL = 160 μL. Diluting each dNTP stock into this final volume gives

$$10 \text{ m}M \text{ each dNTP} \times \frac{40 \text{ μL}}{160 \text{ μL}} = 2.5 \text{ m}M \text{ each dNTP}$$

Therefore, the master mix contains 2.5 m*M* of each dNTP.

Solution 8.16(b) The total concentration of dNTP in the master mix is 10 m*M* (2.5 m*M* dATP + 2.5 m*M* dCTP + 2.5 m*M* dGTP + 2.5 m*M* dTTP = 10 m*M* total dNTP). Dilution of 4 μL of the dNTP master mix into a 50-μL PCR reaction yields the following concentration:

$$10 \text{ m}M \text{ total dNTP} \times \frac{4 \text{ μL}}{50 \text{ μL}} \times \frac{1 \times 10^3 \text{ μ}M}{\text{m}M} = 800 \text{ μ}M \text{ total dNTP}$$

Therefore, the 50-μL PCR reaction contains a total dNTP concentration of 800 μ*M*.

Problem 8.17 A 50-μL PCR reaction is designed to produce a 300-bp amplified product. The reaction contains 800 μ*M* total dNTP. The primers used for the amplification are an 18-mer and a 22-mer. Assume that neither the primers nor any other reagent in the reaction are limiting and that the reaction is 100% efficient. (**a**) With a dNTP concentration of 800 μ*M* dNTP, how much DNA can be produced in this reaction? (**b**) How many copies of the 300-bp fragment could be theoretically produced? (**c**) How much primer is required to synthesize the number of calculated copies of the 300-bp fragment?

Solution 8.17(a) The first step is to calculate the number of micromoles of dNTP in the reaction. (800 μ*M* dNTP is equivalent to 800 μmol dNTP/L.)

$$\frac{800 \text{ μmol dNTP}}{\text{L}} \times \frac{1 \text{ L}}{1 \times 10^6 \text{ μL}} \times 50 \text{ μL} = 0.04 \text{ μmol dNTP}$$

Therefore, the reaction contains 0.04 μmol of dNTP.

A dNTP is a triphosphate. When added onto an extending DNA strand by the action of a DNA polymerase, a diphosphate is released. The nucleotide (nt), as incorporated into a strand of DNA, has a molecular weight of 330 g/mol. Therefore, the number of micromoles of dNTP is equal to the number of micromoles of nucleotide of the form incorporated into DNA. The amount of DNA that can be made from 0.04 μmol of nucleotide is calculated by multiplying this amount by the molecular weight of a single nucleotide. Multiplication by a conversion factor allows expression of the result in micrograms.

$$0.04 \ \mu\text{mol nt} \times \frac{330 \ \text{g/mol}}{\text{nt}} \times \frac{1 \ \text{mol}}{1 \times 10^{6} \ \mu\text{mol}} \times \frac{1 \times 10^{6} \ \mu\text{g}}{\text{g}} = 13.2 \ \mu\text{g}$$

Therefore, 0.04 μmol of dNTP can theoretically be converted into 13.2 μg of DNA.

Solution 8.17(b) A 300-bp PCR fragment, since it is double-stranded, consists of 600 nucleotides. Each PCR fragment was made by extension of annealed primers. The primers in this reaction are 18 and 22 nucleotides in length. Forty nucleotides (18 nt + 22 nt = 40 nt) of each PCR fragment, therefore, are made up from primer. Another 560 nucleotides (600 nt – 40 nt from the primer) of each PCR fragment are made from the dNTP pool. The 560 nucleotides represent 1 copy of the target sequence. Their weight is calculated as follows:

$$\frac{560 \ \text{nt}}{\text{copy}} \times \frac{330 \ \text{g}}{\text{mol nt}} \times \frac{1 \ \text{mol nt}}{6.023 \times 10^{23} \ \text{nt}} \times \frac{1 \times 10^{6} \ \mu\text{g}}{\text{g}} = \frac{3.1 \times 10^{-13} \ \mu\text{g}}{\text{copy}}$$

Therefore, each copy of PCR fragment contains 3.1×10^{-13} μg of nucleotide contributed from the dNTP pool. Multiplying this value by the amount of DNA that can be made from the dNTPs (13.2 μg, as calculated in Solution 8.17a) gives the following result:

$$\frac{1 \ \text{copy}}{3.1 \times 10^{-13} \ \mu\text{g}} \times 13.2 \ \mu\text{g} = 4.3 \times 10^{13} \ \text{copies}$$

Therefore, the dNTPs in this reaction can synthesize 4.3×10^{13} copies of the 300-bp PCR fragment.

Solution 8.17(c) Since the primers are incorporated into each copy of the PCR product, the micromoles of primers needed to synthesize 4.3×10^{13} copies is equivalent to the number of micromoles of product. To solve this problem, then, the number of micromoles of 300-bp product must first be determined:

$$4.3 \times 10^{13} \ \text{copies} \times \frac{1 \ \text{mol}}{6.023 \times 10^{23} \ \text{copies}} \times \frac{1 \times 10^{6} \ \mu\text{mol}}{\text{mol}} = 7.1 \times 10^{-5} \ \mu\text{mol 300 - bp product}$$

Therefore, 7.1×10^{-5} μmol of primer in the 50-μL reaction are needed. To convert this value to a micromolar concentration, the quantity is expressed as an amount per liter:

$$\frac{7.1 \times 10^{-5}\ \mu\text{mol primer}}{50\ \mu\text{L}} \times \frac{1 \times 10^{6}\ \mu\text{L}}{\text{L}} = \frac{1.4\ \mu\text{mol primer}}{\text{L}} = 1.4\ \mu\text{M primer}$$

Therefore, each primer must be present in the reaction at a concentration of 1.4 μ*M* to support the synthesis of 4.3×10^{13} copies of 300-bp PCR product when the total dNTP concentration is 800 μ*M*.

DNA Polymerase

A number of DNA polymerases isolated from different thermophilic bacteria have been successfully used for PCR amplification. Although they may have varying characteristics as to exonuclease activity, processivity, fidelity, and specific activity, they have been chosen as suitable for PCR because they are thermally stable; they retain their activity even after repeated exposure to the high temperatures used to denature the template nucleic acid.

Problem 8.18 Two units of *Taq* DNA polymerase are added to a 100-μL PCR reaction. The enzyme has a specific activity of 292,000 units/mg and a molecular weight of 94 kilodaltons (kDa). How many molecules of the enzyme are in the reaction?

Solution 8.18 Solving this problem requires the use of Avogadro's number (6.023×10^{23} molecules/mol). A molecular weight of 94 kDa is equivalent to 94,000 g/mol. These values are incorporated into the following equation to give the number of enzyme molecules:

$$2\ \text{units enzyme} \times \frac{\text{mg}}{292{,}000\ \text{units enzyme}} \times \frac{\text{g}}{1000\ \text{mg}} \times \frac{\text{mol}}{94{,}000\ \text{g}}$$

$$\times \frac{6.023 \times 10^{23}\ \text{molecules}}{\text{mol}} = x\ \text{molecules}$$

Simplifying the equation gives the following result:

$$\frac{1.2 \times 10^{24}\ \text{molecules}}{2.7 \times 10^{13}} = 4.4 \times 10^{10}\ \text{molecules}$$

Therefore, the reaction contains 4.4×10^{10} molecules of *Taq* DNA polymerase.

Calculating DNA Polymerase's Error Rate
All DNA polymerases, including those used for PCR, display a certain misincorporation rate by which a base other than that complementary to the template is added to the 3′ end of the strand being extended. Most DNA polymerases have a 3′ to 5′ exonuclease "proofreading" function that removes incorrect bases from the 3′ end of the extending strand. Once the incorrect base has been removed, the polymerase activity of the enzyme adds the correct base and extension continues.

Hayashi (1994) describes a method for calculating the fraction of product fragments having the correct sequence after a PCR reaction. It is given by the following equation yielding the percent of PCR fragments containing no errors from misincorporation:

$$F(n) = \frac{[1 + E(P)]^n}{(E+1)^n} \times 100$$

where $F(n)$ = fraction of strands with correct sequence after n cycles

E = efficiency of amplification

n = number of cycles

P = probability of producing error-free PCR fragments in one cycle of amplification in a no-hit Poisson distribution, where λ in the Poisson equation, $P = e^{-\lambda}\lambda^r/r!$ is equal to the DNA polymerase error rate multiplied by the length of the amplicon in nucleotides

Calculating probability by Poisson distribution was described in Chapter 3 for analysis of the fluctuation test. It utilizes the general formula

$$P = \frac{e^{-\lambda}\lambda^r}{r!}$$

where e is the base of the natural logarithm system and r designates the number of "hits." $r!$ (r factorial) designates the product of each positive integer from r down to and including 1. For the zero ("no hit" or no errors incorporated) case, $r = 0$ and the Poisson equation becomes

$$P = \frac{e^{-\lambda}\lambda^0}{0!}$$

When solving P for the no-hit case, remember first that 0! is equal to 1 and second that any number raised to the exponent 0 is equal to 1. The no-hit case is then

$$P = \frac{e^{-\lambda}(1)}{1} = e^{-\lambda}$$

Problem 8.19 Under a defined set of reaction conditions, *Taq* DNA polymerase is found to have an error rate of 2×10^{-5} errors per nucleotide polymerized. An amplicon of 500 bp is synthesized in the reaction. The reaction has an overall amplification efficiency of 0.85. After 25 cycles, what percent of the fragments will have the same sequence as that of the starting template (i.e., will not contain errors introduced by the DNA polymerase)?

Solution 8.19 The Poisson distribution probability for the no-hit case is given by the following equation:

$$P = \frac{e^{-\lambda}\lambda^0}{0!}$$

For this problem, λ is equal to the error rate multiplied by the amplicon size, given by the following equation:

$$\lambda = \left(2 \times 10^{-5}\right)(500) = 0.01$$

Substituting this λ value into the Poisson equation gives the following relationship:

$$P = e^{-0.01} = 0.99$$

[To find $e^{-0.01}$ on a calculator, enter 0.01, change the sign by pressing the **+/–** key, then press the e^x key (on some calculators, it may be necessary first to press the **SHIFT** key to gain access to the e^x function.)]

Bringing the value for P into the equation for calculating the percentage of PCR fragments with no errors gives the following equation.

$$F(n) = \frac{[1 + E(P)]^n}{(E+1)^n} \times 100$$

$$F(n) = \frac{[1 + 0.85\ (0.99)]^{25}}{(0.85+1)^{25}} \times 100$$

Simplifying the equation gives the following relationship:

$$F(n) = \frac{[1.84]^{25}}{(1.85)^{25}} \times 100$$

Both the numerator and the denominator must be raised to the 25th power. To determine $[1.84]^{25}$ on a calculator, enter 1.84, press the $\boldsymbol{x^y}$ key, enter 25, then press the = key.

The equation then gives the following result:

$$F(n) = \frac{4.17 \times 10^6}{4.78 \times 10^6} \times 100 = 87.2$$

Therefore, after 25 cycles, 87.2% of the amplicons will have the correct sequence.

Quantitative PCR

The ability to measure minute amounts of a specific nucleic acid can be crucial to obtaining a thorough understanding of the processes involved in gene expression and infection. Because of the PCR's unique sensitivity and because of its capacity to amplify as few as only several molecules to greater than a million-fold, it has been exploited for use in nucleic acid quantitation. No other technique can be used to detect so small an amount of target so efficiently or so quickly. Although a number of different schemes have been devised to employ the PCR as a quantitative tool, the technique called **competitive PCR** has emerged as the most reliable, non-instrument-based approach.

In a competitive PCR assay, a series of replicate reaction tubes is prepared in which two templates are coamplified. One template, the target sequence being quantified (its quantity is unknown), is added as an extract in identical volumes to each tube. Depending on the application, this target template might be RNA transcribed from an induced gene of interest or an RNA present in biological tissue as a consequence of viral infection. The other template is a competitor added to each PCR reaction as a dilution series in an increasing and known amount. The two templates, the target and the competitor, to ensure that they will be amplified with the same efficiency, have similar but distinguishable lengths, similar sequence, and the same primer-annealing sequences. If RNA is the nucleic acid to be quantified, then the competitor template should also be RNA. The RNAs must first be reverse transcribed into cDNA prior to PCR amplification. (The technique used to prepare cDNA from RNA followed by PCR amplification is termed **RT-PCR**.) During reverse transcription and PCR, the competitor template competes with the target sequence for the primers and reaction components.

Following amplification, the reaction products are separated by gel electrophoresis. The amount of product in each band can be determined by several different methods. If a radioactively labeled dNTP is included in the PCR reaction, autoradiography or liquid scintillation spectrometry can be used to detect the relative amounts of product. Staining with ethidium bromide and gel photography can be used in conjunction with densitometric scanning to quantitate products. If fluorescent dye–labeled primers are used, the products can be quantitated on an automated DNA sequencer.

If quantitation is done either by assessing radioisotope incorporation or by fluorescent/densitometric scanning of an ethidium bromide–stained gel, the quantities obtained for the smaller fragments must be corrected to reflect actual molar amounts. For example, smaller DNA fragments will incorporate less ethidium bromide than larger fragments, even though the actual number of molecules between them may be equivalent. Smaller fragments, therefore, will fluoresce less intensely. To be able to compare their quantities directly, the amount of the smaller fragment should be multiplied by the ratio of bp of the larger fragment to bp of the smaller fragment.

In any PCR reaction, the amount of PCR fragment produced is proportional to the amount of starting template. The ratio of PCR products in the competitive PCR assay reflects the amount of each template initially present in the reaction. However, the total amount of PCR product cannot exceed some maximum value as limited by the amounts of dNTP or active *Taq* polymerase. As the amount of competitor template is increased in the series of reactions, the amount of product from the target template being quantitated will decrease as the competitor effectively competes for the reaction components. In that reaction, where the initial amounts of target and competitor template are equivalent, equal amounts of their products will accumulate. Therefore, at the point where the quantity of the product from the target and the quantity of the product from the competitor template are equal (the **equivalence point**), the starting concentration of target prior to RT-PCR is equal to the known amount of starting competitor template in that reaction.

It is unlikely that, by chance, any one reaction in the series will produce exactly equivalent amounts of product. The equivalence point, therefore, must be determined by constructing a plot relating the logarithm of the ratio of the amount of competitor PCR product/amount of target PCR product to the logarithm of the amount of initial competitor template. Although such a plot may suggest a straight line, regression analysis will need to be conducted to estimate the line of best fit through the data points. The equivalence point on the regression line is equal to 0 on the *y* axis. (At the equivalence point, the ratio of competitor/target is equal to 1. The logarithm of 1 is 0.) Regression will be described more fully in the following problem.

Problem 8.20 Ten reaction tubes are prepared for the quantitiation of an RNA present in an extract from viral-infected cells. Ten μL of extract are placed in each of the 10 tubes. Into the reaction tubes are placed a competitor RNA having the same primer-annealing sites as the target RNA and a similar base sequence. The competitor template is added to the 10 tubes in the following amounts: 0, 2, 10, 20, 50, 100, 250, 500, 1000, and 5000 copies. Each reaction is treated with reverse transcriptase to make cDNA. The reactions are then amplified by PCR. RT-PCR of the target RNA yields a fragment 250 bp in length, while that of the competitor template produces a 200-bp fragment. Following amplification, the products are run on a 2% agarose gel in the presence of 0.5 μg/mL ethidium bromide. A photograph of the gel is taken under UV light, and the fluorescence intensity of each band is quantitated by densitometric scanning. The results shown in Figure 8.1 are obtained.

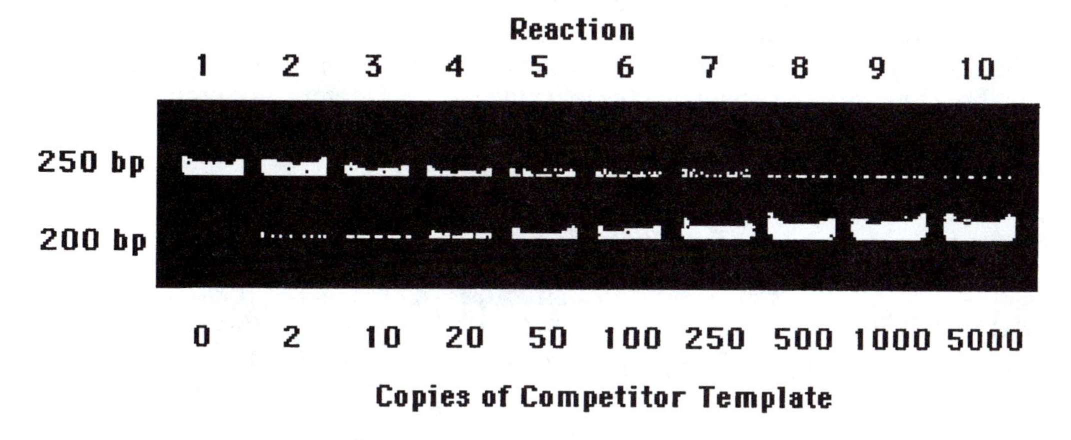

Figure 8.1 Agarose gel of competitive PCR assay. The target yields a 250-bp product. The competitive template yields a 200-bp product.

Densitometric scanning gives the values presented in Table 8.3.

Reaction Tube	Copies of Competitor Added	Scanning Density of Competitor RT-PCR Product	Scanning Density of Target RT-PCR Product
1	0	0	24,822
2	2	3,298	28,437
3	10	8,686	18,218
4	20	7,761	16,872
5	50	14,590	13,520
6	100	16,477	11,212
7	250	21,994	7,115
8	500	25,050	5,173
9	1000	26,598	5,091
10	5000	31,141	1,951

Table 8.3 Scanning data for the gel photograph in Figure 8.1.

Solution 8.20 The PCR product generated from the competitor RNA is smaller than that generated by the target RNA. When run on an agarose gel and stained with ethidium bromide, smaller fragments will stain with less intensity than larger fragments, even though their molar amounts may be equivalent, because the shorter molecules will bind less ethidium bromide. To be able to directly compare the amount of competitor PCR product to that of the target, the scanning density values of the competitor must be corrected by multiplying them by the ratio of the size of the larger fragment over the smaller fragment. For this experiment, the product generated by the target RNA is 250 bp. That generated by the competitor RNA is 200 bp. Therefore, the scanning density values for each competitor product should be multiplied by 250/200. These values are entered in Table 8.4.

Reaction Tube	Copies of Competitor Added	Scanning Density of Competitor RT-PCR Product	Corrected Competitor PCR Product Scanning Density
1	0	0	0
2	2	3,298	4,123
3	10	8,686	10,858
4	20	7,761	9,701
5	50	14,590	18,238
6	100	16,477	20,596
7	250	21,994	27,492
8	500	25,050	31,312
9	1000	26,598	33,247
10	5000	31,141	38,926

Table 8.4 Corrected scanning density values for the 200-bp competitor RNA PCR product for Problem 8.20. Each value is multiplied by 250/200 (bp size of target PCR product/bp size of competitor RNA PCR product). For example, for reaction 2, $3298 \times 250/200 = 4123$.

The *y* axis for our plot will be the log of the corrected competitor PCR product scanning denisity/target PCR product scanning density. Table 8.5 presents these values.

Reaction Tube	Corrected Competitor PCR Product Scanning Density	Scanning Density of Target RT-PCR Product	Competitor Divided by Target	Log of Competitor Divided by Target
1	0	24,822	—	—
2	4,123	28,437	0.145	–0.839
3	10,858	18,218	0.596	–0.225
4	9,701	16,872	0.575	–0.240
5	18,238	13,520	1.349	0.130
6	20,596	11,212	1.837	0.264
7	27,492	7,115	3.864	0.587
8	31,312	5,173	6.053	0.782
9	33,247	5,091	6.531	0.815
10	38,926	1,951	19.953	1.300

Table 8.5 Competitor PCR product scanning density divided by the target PCR product scanning density is given in the fourth column. The log of these values (y axis values) are shown in the fifth column.

The x axis is the log of the number of copies of competitor RNA added to each PCR reaction in the assay. Table 8.6 provides these values.

Reaction Tube	Copies of Competitor Added	Log of Competitor Copies
1	0	—
2	2	0.3
3	10	1.0
4	20	1.3
5	50	1.7
6	100	2.0
7	250	2.4
8	500	2.7
9	1000	3.0
10	5000	3.7

Table 8.6 Log values for the number of copies of competitor RNA placed in each tube of the quantitative PCR assay for Problem 8.20. For example, for reaction 2, in which two copies of the competitive template RNA were placed in the reaction, the log of 2 is 0.3.

A plot can now be generated using log copies of competitor RNA on the x axis and log competitor/target on the y axis (Figure 8.2).

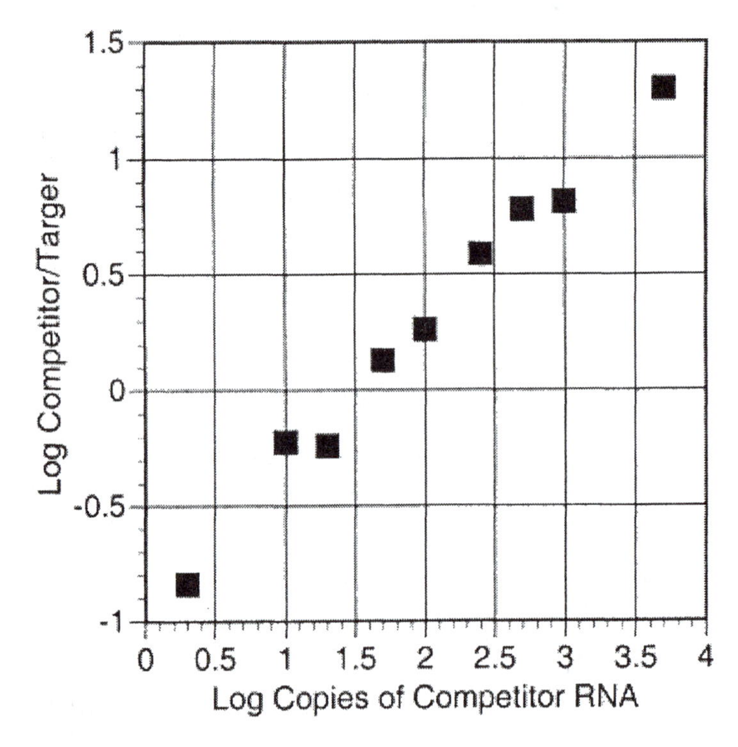

Figure 8.2 Scatter plot of log competitor/target vs. log copies of competitor RNA.

The points on the graph in Figure 8.2 suggest a straight-line relationship, but not all points fall exactly on a straight line. Such a graph is called a **scatter plot** because the points are scattered across the grid. The object then becomes to draw a line through the scatter plot that will best fit the data. Such a best-fit line can be calculated by a technique called **regression analysis** (also called **linear regression** or the **method of least squares**). The values on the x axis are called the **independent** or **predictor variables**. They are chosen by the experimenter and are therefore not random. (In this example, the log of the number of copies of competitor RNA placed in the series of competitive PCR reactions is the independent variable.) The values on the y axis are the **dependent** or **response variables**. (In this example, the log of the corrected fluorescence of the competitor PCR product gel band divided by the fluoresence of the target PCR product band is the dependent variable.) Regression analysis creates a best-fit line that minimizes the distance of all the data points from the line. The regression line drawn to best fit the scatter plot has the formula

$$y = mx + b$$

This equation, in fact, describes any straight line in two-dimensional space. The y term is the variable in the vertical axis, m is the slope of the line, x is the variable on the horizontal axis, and b is the value of y where the line crosses the vertical axis (called the **intercept**).

For the best-fit regression line,

$$b = \bar{y} - m\bar{x}$$

where $\bar{y}$ is the mean y value, $\bar{x}$ is the mean x value, and

$$m = \frac{\sum_{i=1}^{n}(x_i - \bar{x})(y_i - \bar{y})}{\sum^{n}(x_i - \bar{x})^2}$$

where n designates the number of reactions (the number of data points), i refers to a specific PCR reaction in the series of competitive reactions ($i = 1$ to n), and the Greek letter Σ (sigma) is used as summation notation to indicate the sum of all terms in the series for each value of i.

We can now use our data to determine the regression line. The first step is to calculate the mean values for x_i and y_i. This is done by adding all the values for x_i and y_i (Table 8.7) and then dividing each sum by the total number of samples used to obtain that sum. (The values for reaction 1, in which no competitor template was added will not be used for these calculations because this reaction provides no useful information other than to give us peace of mind that a 200-bp product would not be produced in the absence of competitor template.)

Reaction	x_i [Log Copies of Competitor Added (See Table 8.6)]	y_i [Log of Competitor Divided by Target (See Table 8.5)]
2	0.3	–0.839
3	1.0	–0.225
4	1.3	–0.240
5	1.7	0.130
6	2.0	0.264
7	2.4	0.587
8	2.7	0.782
9	3.0	0.815
10	3.7	1.300
Total	**18.1**	**2.574**

Table 8.7 Calculating the sum of all x and y values.

The mean value for x is calculated as follows:

$$\bar{x} = \frac{18.1}{9} = 2.0$$

The mean value for y is calculated as follows:

$$\bar{y} = \frac{2.574}{9} = 0.286$$

The values for m and b in the equation for the regression line can now be calculated. Table 8.8 gives the values necessary for these calculations.

x_i	y_i	$(x_i - \bar{x})$	$(y_I - \bar{y})$	$(x_i - \bar{x})^2$	$(y_I - \bar{y})^2$	$(x_i - \bar{x}) \times (y_I - \bar{y})$
0.3	–0.839	–1.7	–1.125	2.89	1.266	1.913
1.0	–0.225	–1.0	–0.511	1.00	0.261	0.511
1.3	–0.240	–0.7	–0.526	0.49	0.277	0.368
1.7	0.130	–0.3	–0.156	0.09	0.024	0.047
2.0	0.264	0.0	–0.022	0.00	0.000	0.000
2.4	0.587	0.4	0.301	0.16	0.091	0.120
2.7	0.782	0.7	0.496	0.49	0.246	0.347
3.0	0.815	1.0	0.529	1.00	0.280	0.529
3.7	1.300	1.7	1.014	2.89	1.028	1.724
18.1	**2.574**			**9.01**	**3.473**	**5.559**

Table 8.8 Calculation of values necessary to determine *m* and *b* for the regression line of best fit.

From Table 8.8, we have the following values:

$$\sum^{9}(x_i - \bar{x})^2 = 9.01$$

and

$$\sum_{i=1}^{9}(x_i - \bar{x})(y_i - \bar{y}) = 5.559$$

With these values, we can calculate *m*:

$$m = \frac{\sum_{i=1}^{9}(x_i - \bar{x})(y_i - \bar{y})}{\sum^{9}(x_i - \bar{x})^2} = \frac{5.559}{9.01} = 0.617$$

Knowing the value for *m*, the value for *b* can be calculated:

$$\begin{aligned} b &= \bar{y} - m\bar{x} \\ &= 0.286 - 0.617(2.0) \\ &= 0.286 - 1.234 = -0.948 \end{aligned}$$

Therefore, substituting the m and b values into the equation for a straight line gives

$$y = 0.617x + (-0.948)$$

A regression line can now be plotted. Since a line is defined by two points, we need to determine two values for the equation $y = mx + b$. The regression line always passes through the point ($\bar{x}$, $\bar{y}$), in this example (2.0, 0.286). For a second point, let's set x equal to 3.5. (This choice is arbitrary; any x value within the experimental range will do.) Solving for y using our regression line equation gives

$$y = 0.617(3.5) + (-0.948)$$
$$y = 2.160 - 0.948$$
$$y = 1.212$$

Plotting these two points, (2.0, 0.286) and (3.5, 1.212), and drawing a straight line between them gives the line of best fit (Figure 8.3).

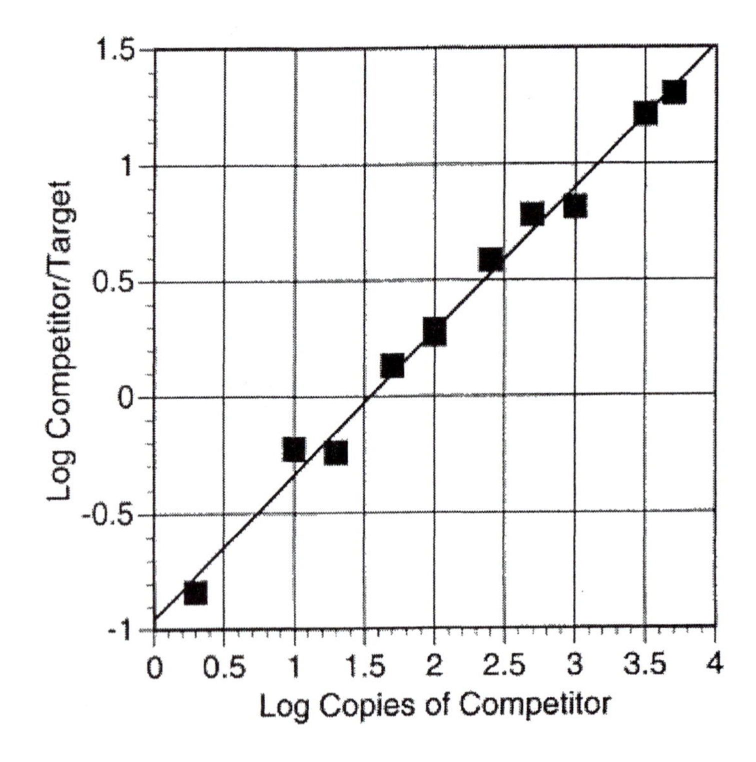

Figure 8.3 Line of best fit using the regression equation $y = 0.617x + (-0.948)$.

The equivalence point can be calculated using our equation for the regression line and setting y equal to 0, as shown in the following equation:

$$y = 0.617x + (-0.948)$$
$$0 = 0.617x - 0.948$$
$$0.948 = 0.617x$$
$$\frac{0.948}{0.617} = x$$
$$x = 1.536$$

Since this number represents the $\log_{10}$ value of the copies of competitor, we must find its anitlog. On a calculator, this is done by entering 1.536 and then pressing the $\mathbf{10^x}$ key.

$$\text{antilog}_{10} 1.536 = 34$$

Therefore, the aliquot of cell extract contained 34 copies of mRNA template.

How closely the regression line fits the data (or how closely the data approaches our best-fit regression line) can be determined by calculating the **sum of the squared errors**, a measure of the total spread of the y values from the regression line. To do this, we will define $\hat{y}_i$ as the value for y_I that lies on the best-fit regression line for each value of x_i such that

$$\hat{y}_i = a + bx_i$$

Error is defined as the amount y_i deviates from $\hat{y}_i$ and is represented by the vertical distance from each data point to the regression line (Figure 8.4). **Regression** measures the difference between the $\hat{y}_i$ value and the mean for y ($\bar{y}$).

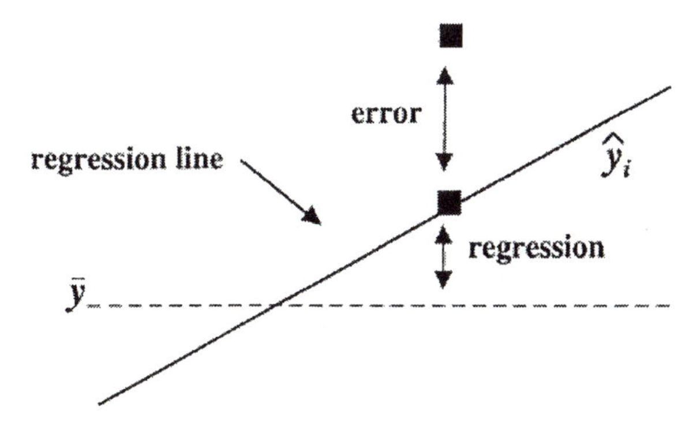

Figure 8.4 Error is the amount y_i deviates from $\hat{y}_i$.

The measure of how well the regression line fits the plot is given by the **squared correlation, R^2**. It is the proportion of the total $(y - \bar{y})^2$ values accounted for by the regression, or a measure of how many y_i values fall directly on the best-fit regression line. It is given by the following equation using the square of the regression values for each point:

$$R^2 = \frac{\sum_{i=1}^{n} (\hat{y}_i - \bar{y})^2}{\sum_{i=1}^{n} (y_i - \bar{y})^2}$$

R^2 can also be calculated using the error values for each point. It is found using the following formula:

$$R^2 = 1 - \frac{\sum_{i=1}^{n} (y_i - \hat{y})^2}{\sum_{i=1}^{n} (y_i - \bar{y})^2}$$

Table 8.9 provides the values necessary to calculate R^2 by either method. Previously, it was calculated that the mean for x is 2.0 and the mean for y is 0.286. Values for $\hat{y}_i$ are calculated using the equation

$$\hat{y}_i = mx_i + b$$

where b is equal to –0.948 (page 179) and m is equal to 0.617 (page 179).

			Regression		Error	
x_i	y_i	$\hat{y}_i$	$(\hat{y}_i - \bar{y})$	$(\hat{y}_i - \bar{y})^2$	$(y_i - \hat{y}_i)$	$(y_i - \hat{y}_i)^2$
0.3	–0.839	–0.763	–1.049	1.100	–0.076	0.006
1.0	–0.225	–0.331	–0.617	0.381	0.106	0.011
1.3	–0.240	–0.146	–0.432	0.187	–0.094	0.009
1.7	0.130	0.101	–0.185	0.034	0.029	0.001
2.0	0.264	0.286	0.000	0.000	–0.022	0.000
2.4	0.587	0.533	0.247	0.061	0.054	0.003
2.7	0.782	0.718	0.432	0.187	0.064	0.004
3.0	0.815	0.903	0.617	0.381	–0.088	0.008
3.7	1.300	1.335	1.049	1.100	–0.035	0.001
Totals				**3.431**		**0.043**

Table 8.9 Values for calculating R^2.

From Table 8.9, the following values are derived:

$$\sum_{i=1}^{9}(\hat{y}_i - \bar{y})^2 = 3.431$$

$$\sum_{i=1}^{9}(y_i - \hat{y}_i)^2 = 0.043$$

The sum of $(y_i - \bar{y})^2$ was calculated in Table 8.8 and was determined to be 3.473. Using the first equation for calculating R^2 gives the following result:

$$R^2 = \frac{\sum_{i=1}^{n}(\hat{y}_i - \bar{y})^2}{\sum_{i=1}^{n}(y_i - \bar{y})^2}$$

$$R^2 = \frac{3.431}{3.473} = 0.988$$

The same result is obtained using the alternate method:

$$R^2 = 1 - \frac{\sum_{i=1}^{n}(y_i - \hat{y})^2}{\sum_{i=1}^{n}(y_i - \bar{y})^2}$$

$$R^2 = 1 - \frac{0.043}{3.473}$$

$$= 1 - 0.012 = 0.988$$

Therefore, the plotted regression line has an R^2 value of 0.988.

The closer the R^2 value is to 1.0, the better the fit of the regression line. An R^2 value of 1.0 corresponds to a perfect fit; all values for y lie on the regression line. In real-world experimentation, R^2 will be less than 1.0. R^2 will equal zero if x_i and y_i are completely independent; that is, in knowing a value for x_i, you will in no way be able to predict a corresponding value for y_i. An R^2 value of 0 might be expected if the scatter plot is completely random, that is, if there are data points all over the graph, with no relationship between x and y.

An alternate measure of how well the regression line fits the plot of experimental data is given by the **correlation coefficient**, ***r***, represented by the equation

$$r = (\text{sign of } m)\sqrt{R^2}$$

where r is a positive value if the regression line goes up to the right (has a positive slope) and negative if the line goes down to the right (has a negative slope). If a relationship exists such that if one value (x) changes, then the other (y) does so in a related manner, the two values are said to be correlated. The value of r ranges from –1 to 1, with 1 indicating a perfect linear relationship between x and y and a positive slope, –1 indicating a perfect linear relationship with a negative slope. Zero indicates no linear relationship.

For this example, m is a positive value (+0.617; see page 179) and the equation for r is

$$r = \sqrt{R^2}$$

$$= \sqrt{0.988} = 0.994$$

Therefore, the regression line has a correlation coefficient, r, of 0.994.

References and Further Reading

Breslauer, Kenneth J., Ronald Frank, Helmut Blöcker, and Luis A. Marky (1986). Predicting DNA duplex stability from the base sequence. *Proc. Natl. Acad. Sci. USA* 83:3746–3750.

Gilliland, Gary, Steven Perrin, Kerry Blanchard, and H. Franklin Bunn (1990). Analysis of cytokine mRNA and DNA: detection and quantitation by competitive polymerase chain reaction. *Proc. Natl. Acad. Sci. USA* 87:2725–2729.

Hayashi, Kenshi (1994). Manipulation of DNA by PCR. In *The Polymerase Chain Reaction* (K. Mullis, F. Ferré. R.A. Gibbs, eds.). Birkhäuser, Boston, pp. 3–13.

Piatak, M., Jr., Ka-Cheung Luk, Bill Williams, and Jeffrey D. Lifson (1993). Quantitative competitive polymerase chain reaction for accurate quantitation of HIV DNA and RNA species. *BioTechniques* 14:70–80.

Quartin, Robin S., and James G. Wetmur (1989). Effect of ionic strength on the hybridization of oligodeoxynucleotides with reduced charge due to methylphosphonate linkages to unmodified oligodeoxynucleotides containing the complementary sequence. *Biochemistry* 28:1040–1047.

Wetmur, James G., and John J. Sninsky (1995). Nucleic acid hybridization and unconventional bases. In *PCR Strategies* (Innis, M.A., D.H. Gelfand, and J.J. Sninsky, eds.). Academic Press, San Diego, CA, pp. 69–83.

Recombinant DNA 9

Introduction

The development of **recombinant DNA** technology (also called **gene cloning** or **gene splicing**), the methodology used to join DNA from different biological sources for the purpose of determining the sequence or for manipulating its expression, spawned an era of gene discovery and launched an industry. In its simplest terms, gene cloning is achieved in four steps.

1. A particular DNA is first fragmented into smaller pieces.
2. A specific segment of the fragmented DNA is then joined to another DNA, termed a **vector**, capable of intracellular replication.
3. The recombinant molecule is introduced into a host cell, where it can replicate, producing multiple copies of the cloned DNA fragment.
4. Cells containing recombinant clones are identified and propagated.

Restriction Endonucleases

Restriction endonucleases are enzymes that recognize a specific DNA sequence, called a **restriction site**, and cleave the DNA within or adjacent to that site. For example, the restriction endonuclease *Eco*R I, isolated from the bacterium *Escherichia coli*, recognizes the following sequence:

5′ GAATTC 3′
3′ CTTAAG 5′

It cleaves the DNA between the G and A on each strand, producing 5′ overhangs of four nucleotides, as shown here:

5′ ---------G 3′ and 5′ AATTC-------- 3′
3′ ---------CTTAA 5′ 3′ G--------5′

The termini produced by *EcoR* I, since they are complementary at their single-stranded overhangs, are said to be **cohesive** or "**sticky**."

A number of restriction enzymes have been isolated from a variety of microbial sources. Recognition sites for specific enzymes range in size from 4 to 13 base pairs, and, for most restriction enzymes used in gene cloning, are palindromes; sequences that read in the 5′-to-3′ direction on one strand, are the same as those in the 5′-to-3′ direction on the opposite strand. Restriction enzyme activity is usually expressed in terms of units, in which one unit is that amount of enzyme that will cleave all specific sites in 1 μg of a particular DNA sample (usually lambda DNA) in one hour at 37°C.

Problem 9.1 A stock of human genomic DNA has a concentration of 275 μg/mL. You wish to cut 5 μg with 10 units of the restriction endonuclease *Hin*d III. The *Hin*d III enzyme has a concentration of 2500 units/mL. What volumes of DNA and enzyme should be used?

Solution 9.1 Restriction digests are typically performed in small volumes (20 to 100 μL). Such reactions are prepared with a micropipettor delivering microliter amounts. The concentrations of both the human DNA and the enzyme stocks are given as a quantity per milliliter. A conversion factor, therefore, must be used to convert milliliters to microliters. The volumes needed for each reaction component can be written as follows.

The amount of input DNA is calculated as follows:

$$\frac{275\ \mu\text{g DNA}}{\text{mL}} \times \frac{1\ \text{mL}}{1000\ \mu\text{L}} \times x\ \mu\text{L} = 5\ \mu\text{g DNA}$$

Solving the equation for x yields the following result:

$$\frac{(275\ \mu\text{g DNA})x}{1000} = 5\ \mu\text{g DNA}$$

$$(275\ \mu\text{g DNA})x = 5000\ \mu\text{g DNA}$$

$$x = \frac{5000\ \mu\text{g DNA}}{275\ \mu\text{g DNA}} = 18.2$$

Therefore, 18.2 μL of the human DNA stock contains 5 μg of DNA.

The amount of *Hin*d III enzyme is calculated as follows:

$$\frac{2500\ \text{units}}{\text{mL}} \times \frac{1\ \text{mL}}{1000\ \mu\text{L}} \times x\ \mu\text{L} = 10\ \text{units}$$

Solving this equation for x yields the following result:

$$\frac{(2500 \text{ units})x}{1000} = 10 \text{ units}$$

$$(2500 \text{ units})x = 10{,}000 \text{ units}$$

$$x = \frac{10{,}000 \text{ units}}{2500 \text{ units}} = 4$$

Therefore, 4 μL of the enzyme stock is needed to deliver 10 units.

The Frequency of Restriction Endonuclease Cut Sites

The frequency with which a particular restriction site occurs in any DNA depends on its base composition and the length of the recognition site. For example, in mammalian genomes, a G base follows a C in a much lower frequency than would be expected by chance. Therefore, a restriction endonuclease such as *Nru* I, which recognizes the sequence TCGCGA, cuts mammalian DNA less frequently than it does DNA from bacterial sources having a more random distribution of bases. Also, as might be expected, restriction sites that are 4 bp in length will occur more frequently than restriction sites that are 6 bp in length. The average size of a restriction fragment produced from a random sequence by a particular endonuclease can be estimated by the method illustrated in the following problem.

Problem 9.2 The restriction endonuclease *Hin*d III recognizes the sequence AAGCTT. If genomic DNA of random sequence is cleaved with *Hin*d III, what will be the average size of the fragments produced?

Solution 9.2 The chance that any one base, A, C, G, or T, will occur at any particular position in DNA of random sequence is 1 in 4. The number of cutting sites in random DNA can be estimated by raising 1/4 to the n [$(1/4)^n$], where the exponent n is the number of base pairs in the recognition sequence. A "6-cutter" such as the restriction enzyme *Hin*d III, which recognizes a 6-bp sequence, will cut at a frequency of $(1/4)^6$:

$$\left(\frac{1}{4}\right)^6 = \frac{1}{4096}$$

Therefore, in DNA of random sequence, a *Hin*d III site might be expected, on average, every 4096 bp along the molecule. (This value reflects only an estimate, and actual sizes of restriction fragments generated from any DNA may vary considerably. For example, bacteriophage lambda DNA, 48,502 bp in length, generates *Hind* III fragments ranging in size from 125 bp to over 23,000 bp.)

Calculating the Amount of Fragment Ends

DNA is digested with restriction enzymes to create fragments for cloning, for gene mapping, or for the production of genetic probes. For some applications, such as fragment end labeling, it is important for the experimenter to estimate the amount of DNA fragment ends. For a given amount of a linear fragment having a certain length, the moles of fragment ends are given by the following equation:

$$\text{moles of DNA ends} = \frac{2 \times (\text{grams of DNA})}{(\text{number of bp}) \times \left(\frac{660 \text{ g/mol}}{\text{bp}}\right)}$$

(There is a 2 in the numerator of this equation because a linear DNA fragment has two ends.)

Problem 9.3 You have 4 μg of a 3400-bp fragment. How many picomoles of ends does this represent?

Solution 9.3 First, calculate the moles of DNA ends and then convert this amount to picomoles. The amount of fragment is expressed in micrograms. Use of the preceding equation requires that the fragment amount be converted to grams. This is done as follows:

$$4 \ \mu\text{g} \times \frac{1 \text{ g}}{1 \times 10^{6} \ \mu\text{g}} = 4 \times 10^{-6} \text{ g}$$

Substituting the given values into the equation for calculating moles of DNA ends yields the following result:

$$\text{moles of DNA ends} = \frac{2 \times (\text{grams of DNA})}{(\text{number of bp}) \times \left(\frac{660 \text{ g/mol}}{\text{bp}}\right)}$$

$$\text{moles of DNA ends} = \frac{2 \times (4 \times 10^{-6} \text{ g})}{(3400 \text{ bp}) \times \left(\frac{660 \text{ g/mol}}{\text{bp}}\right)} = \frac{8 \times 10^{-6} \text{ g}}{2.2 \times 10^{6} \text{ g/mol}}$$

$$= 3.6 \times 10^{-12} \text{ mol}$$

This value is converted to picomoles by multiplying by the conversion factor 1×10^{12} pmol/mol.

$$3.6 \times 10^{-12} \text{ mol} \times \frac{1 \times 10^{12} \text{ pmol}}{\text{mol}} = 3.6 \text{ pmol}$$

Therefore, 4 μg of a 3400-bp fragment are equivalent to 3.6 pmol of DNA ends.

The Amount of Ends Generated by Multiple Cuts

The number of moles of ends created by a restriction digestion of a DNA containing multiple recognition sites can also be determined. For a circular DNA (such as a plasmid or cosmid), the moles of ends created by a restriction digest can be calculated by the following equation:

$$\text{moles ends} = 2 \times (\text{moles DNA}) \times (\text{number of restriction sites})$$

The 2 multiplier at the beginning of this expression is necessary since each restriction enzyme cut generates two ends.

The moles of ends created by multiple-site digestion of a linear DNA molecule are given by the following expression:

$$\text{moles ends} = [2 \times (\text{moles DNA}) \times (\text{number of restriction sites})] + [2 \times (\text{moles DNA})]$$

In this equation, the function at the end of the second expression in brackets is necessary to account for the two ends already on the linear molecule.

Problem 9.4 A 4200-bp plasmid contains five sites for the restriction endonuclease *Eco*R II. If 3 μg are cut with this restriction enzyme, how many picomoles of ends will be produced?

Solution 9.4 Since a circular plasmid is being digested, we will use the following equation:

$$\text{moles ends} = 2 \times (\text{moles DNA}) \times (\text{number of restriction sites})$$

This equation requires that we determine the number of moles of plasmid DNA. To determine that value, the plasmid's molecular weight must first be determined. Moles of plasmid can then be calculated from the amount of plasmid DNA to be digested (3 μg). The molecular weight of the plasmid is calculated as follows:

$$4200 \text{ bp} \times \frac{660 \text{ g/mol}}{\text{bp}} = 2.8 \times 10^{6} \text{ g/mol}$$

The number of moles of plasmid represented by 3 μg is determined by multiplying the amount of plasmid (converted μg to g) by the plasmid's molecular weight:

$$3 \text{ μg} \times \frac{1 \text{ g}}{1 \times 10^{6} \text{ μg}} \times \frac{1 \text{ mol}}{2.8 \times 10^{6} \text{ g}} = \frac{3 \text{ mol}}{2.8 \times 10^{12}} = 1.1 \times 10^{-12} \text{ mol}$$

Therefore, 3 μg of a 4200-bp plasmid are equivalent to 1.1×10^{-12} moles. Placing this value into the equation for moles of ends created from digestion of a circular molecule gives the following result:

$$\text{moles of ends} = 2 \times \left(1.1 \times 10^{-12} \text{ mol}\right) \times 5 = 1.1 \times 10^{-11} \text{ mol}$$

Therefore, 1.1×10^{-11} moles of ends are generated by an *Eco*R II digest of the 4200-bp plasmid. This value is equivalent to 11 picomoles of ends, as shown here:

$$1.1 \times 10^{-11} \text{ mol} \times \frac{1 \times 10^{12} \text{pmol}}{\text{mol}} = 11 \text{ pmol}$$

Problem 9.5 Four μg of a purified 6000-bp linear DNA fragment are to be cut with *Alu* I. There are 12 *Alu* I sites in this fragment. How many picomoles of ends will be created by digesting the fragment with this enzyme?

Solution 9.5 This problem will be approached in the same manner as that followed for Problem 9.4. The molecular weight of the 6000-bp fragment is calculated first:

$$6000 \text{ bp} \times \frac{660 \text{ g/mol}}{\text{bp}} = 4 \times 10^{6} \text{ g/mol}$$

The number of moles of 6000-bp fragment in 4 μg is then determined:

$$4 \text{ μg} \times \frac{1 \text{ g}}{1 \times 10^{6} \text{ μg}} \times \frac{1 \text{ mol}}{4 \times 10^{6} \text{ g}} = \frac{4 \text{ mol}}{4 \times 10^{12}} = 1 \times 10^{-12} \text{ mol DNA}$$

Placing this value into the equation for calculating moles of ends created from digestion of a linear molecule gives the following relationship:

$$\text{moles of ends} = [2 \times (\text{moles DNA}) \times (\text{number of restriction sites})] + [2 \times (\text{moles DNA})]$$

$$\text{moles of ends} = [2 \times (1 \times 10^{-12} \text{ mol}) \times (12)] + [2 \times (1 \times 10^{-12} \text{ mol})]$$

$$= [2.4 \times 10^{-11} \text{ mol}] + [2 \times 10^{-12} \text{ mol}] = 2.6 \times 10^{-11} \text{ mol}$$

This value is converted to picomoles of ends in the following manner:

$$2.6 \times 10^{-11} \text{ mol} \times \frac{1 \times 10^{12} \text{ pmol}}{\text{mol}} = 26 \text{ pmol}$$

Therefore, digesting 4 μg of the 6000-bp fragment with *Alu* I will generate 26 pmol of ends.

Ligation

Once the DNA to be cloned exists as a defined fragment (called the **target** or **insert**), it can be joined to the vector by the process called **ligation**. The tool used for this process is DNA ligase, an enzyme that catalyzes the formation of a phosphodiester bond between juxtaposed 5′-phosphate and 3′-hydroxyl termini in duplex DNA. The desired product in a ligation reaction is a functional hybrid molecule that consists exclusively of the vector plus the insert. However, a number of other events can occur in a ligation reaction, including circularization of the target fragment, ligation of vector without the addition of an insert, ligation of a target molecule to another target molecule, and any number of other possible combinations.

Two vector types are used in recombinant DNA research, circular molecules such as plasmids and cosmids and linear cloning vectors such as those derived from the bacteriophage lambda. In either case, for joining to a target DNA fragment, a vector is first cut with a restriction enzyme that produces compatible ends with those of the target. Circular vectors, therefore, are converted to a linear form prior to ligation to target. The insert fragment is then ligated to the prepared vector to create a recombinant molecule capable of replication once introduced into a host cell.

Dugaiczyk et al. (1975) described the events occurring during ligation of *EcoR* I fragments, observations that can be applied to any ligation reaction. They identified the two factors that play the biggest role in determining the outcome of any particular ligation. First, the readiness with which two DNA molecules join is dependent on the concentration of their ends; the higher the concentration of compatible ends (those capable of being joined), the greater the likelihood that two termini will meet and be ligated. This parameter is designated by the term i and is defined as the *total* concentration of complementary ends in the ligation reaction. Second, the amount of circularization occurring within a ligation reaction is dependent on the parameter j, the concentration of same-molecule ends in close enough proximity to each other that they could potentially and effectively interact. For any DNA fragment, j is a constant value dependent on the fragment's length, not on its concentration in the ligation reaction.

For a linear fragment of duplex DNA with self-complementary, cohesive ends (such as those produced by *EcoR* I), i is given by the formula

$$i = 2N_0M \times 10^{-3} \text{ ends/mL}$$

where N_0 is Avogadro's number (6.023×10^{23}) and M is the molar concentration of the DNA.

Some DNA fragments to be used for cloning are generated by a double digest, say by using both *EcoR* I and *Hin*d III in the same restriction reaction. In such a case, one end of the fragment will be cohesive with other *EcoR* I-generated termini, while the other end of the DNA fragment will be cohesive with other *Hin*d III-generated termini. For such DNA fragments with ends that are not self-complementary, the concentration of each end is given by

$$i = N_0M \times 10^{-3} \text{ ends/mL}$$

Dugaiczyk et al. determined that a value for j, the concentration of one end of a molecule in the immediate vicinity of its other end, can be calculated using the equation

$$j = \left(\frac{3}{2\pi lb}\right)^{\frac{3}{2}} \text{ ends/mL}$$

where l is the length of the DNA fragment, b is the minimal length of DNA that can bend around to form a circle, and π is the number pi, the ratio of the circumference of a circle to its diameter (approximately 3.14, a number often seen when performing calculations relating to measurements of the area of a circle). They specifically determined the value of j for bacteriophage lambda DNA. Lambda DNA is 48,502 bp in length, and they assigned a value of 13.2×10^{-4} cm to l. For lambda DNA in ligation buffer, they assigned a value of 7.7×10^{-6} cm to b. Using these values, the j value for lambda (j_λ) is calculated to be 3.22×10^{11} ends/mL.

The j value for any DNA molecule can be calculated in relation to j_λ by the equation

$$j = j_\lambda \left(\frac{\text{MW}_\lambda}{\text{MW}}\right)^{1.5} \text{ ends/mL}$$

where j_λ is equal to 3.22×10^{11} ends/mL and "MW" represents molecular weight. Since the lambda genome is 48,502 bp in length, its molecular weight can be calculated using the conversion factor 660 g/mol/bp, as shown in the following equation:

$$48{,}502 \text{ bp} \times \frac{660 \text{ g/mol}}{\text{bp}} = 3.2 \times 10^7 \text{ g/mol}$$

Therefore, the molecular weight of lambda (MW_λ) is 3.2×10^7 g/mol.

Under circumstances in which j is equal to i ($j = i$, or $j/i = 1$), the end of any particular DNA molecule is just as likely to join with another molecule as it is to interact with its own opposite end. If j is greater than i ($j > i$), intramolecular ligation events predominate and circles are the primary product. If i is greater than j ($i > j$), intermolecular ligation events are favored and hybrid linear structures predominate.

Ligation Using Lambda-Derived Vectors

The vectors derived from phage lambda carry a cloning site within the linear genome. Cleavage at the cloning site by a restriction enzyme produces two fragments, each having a lambda cohesive end sequence (*cos*) at one end and a restriction enzyme-generated site at the other. Successful ligation of target with lambda vector actually requires three primary events.

1. Joining of the target segment to one of the two lambda fragments
2. Ligation of the other lambda fragment to the other end of the target fragment to create a full-length lambda genome equivalent
3. Joining of the cos ends between other full-length lambda molecules to create a **concatemer**, a long DNA molecule in which a number of lambda genomes are joined in a series, end-to-end. Only from a concatemeric structure can individual lambda genomes be efficiently packaged into a protein coat allowing for subsequent infection and propagation.

In ligation of a fragment to the arms of a λ vector, both ends of the fragment will be compatible with the cloning site of either lambda arm, but each arm has only one compatible end with the insert. For the most efficient ligation, the reaction should contain equimolar amounts of the restriction enzyme-generated ends of each of the three fragments, the λ left arm, the insert, and the λ right arm. However, since there are two fragments of lambda and one insert fragment per recombinant genome, when preparing the ligation reaction, the molar ratio of annealed arms to insert should be 2:1 so that the three molecules have a ratio of 2:1:2 (left arm:insert:right arm). Furthermore, since linear ligation products need to predominate, the concentration of ends (i) and the molecule's length (j) must be manipulated such that j is less than i, thereby favoring the production of linear hybrid molecules over circular ones.

Problem 9.6 You have performed a partial *Sau*3A I digest of human genomic DNA, producing fragments having an average size of 20,000 bp. You will construct a **genomic library** (a collection of recombinant clones representing the entire genetic complement of a human) by ligating these fragments into a λEMBL3 vector digested with *Bam*H I. (*Bam*H I creates ends compatible with those generated by *Sau*3A I; both enzymes produce a GATC 5' overhang.) The λEMBL3 vector is approximately 42,360 bp in length. Removal of the 13,130-bp "stuffer" fragment (the segment between the *Bam*H I sites, which is replaced by the target insert), leaves two arms, which, if annealed, would give a linear genome of 29,230 bp. **(a)** If you wish to prepare a 150-μL ligation reaction, how many micrograms of insert and vector arms should be combined to give optimal results? **(b)** How many micrograms of λEMBL3 vector should be digested with *Bam*H I to yield the desired amount of arms?

Solution 9.6(a) Since we wish to ensure the formation of linear concatemers to optimize the packaging/infection step, j should be less than i. The first step then is to calculate the j values for the λ arms (j_{arms}) and the inserts ($j_{inserts}$). The following equation can be used:

$$j = j_{\lambda}\left(\frac{MW_{\lambda}}{MW}\right)^{1.5} \text{ ends/mL}$$

where j_{λ} is 3.22×10^{11} ends/mL and, as calculated previously, MW_{λ} = 3.2×10^{7} g/mol.

The molecular weight of the λEMBL3-annealed arms is calculated as follows:

$$29{,}230 \text{ bp} \times \frac{660 \text{ g/mol}}{\text{bp}} = 1.93\times10^{7} \text{ g/mol}$$

The j_{arms} value can now be calculated by substituting the proper values into the equation for determining j:

$$j_{arms} = \left(3.22\times10^{11} \text{ ends/mL}\right)\left(\frac{3.2\times10^{7} \text{ g/mol}}{1.93\times10^{7} \text{ g/mol}}\right)^{1.5}$$

$$j_{\text{arms}} = (3.22 \times 10^{11} \text{ ends/mL})\left(\frac{1.8 \times 10^{11}}{8.48 \times 10^{10}}\right)$$

$$j_{\text{arms}} = (3.22 \times 10^{11} \text{ ends/mL})(2.12)$$

$$j_{\text{arms}} = 6.83 \times 10^{11} \text{ ends/mL}$$

Therefore, j for the annealed arms is equal to 6.83×10^{11} ends/mL.

The molecular weight of a 20,000-bp insert is calculated as follows:

$$20{,}000 \text{ bp} \times \frac{660 \text{ g/mol}}{\text{bp}} = 1.3 \times 10^{7} \text{ g/mol}$$

The j_{inserts} value can now be calculated.

$$j_{\text{inserts}} = \left(3.22 \times 10^{11} \text{ ends/mL}\right)\left(\frac{3.2 \times 10^{7} \text{ g/mol}}{1.3 \times 10^{7} \text{ g/mol}}\right)^{1.5}$$

$$j_{\text{inserts}} = \left(3.22 \times 10^{11} \text{ ends/mL}\right)\left(\frac{1.8 \times 10^{11}}{4.7 \times 10^{10}}\right)$$

$$j_{\text{inserts}} = \left(3.22 \times 10^{11} \text{ ends/mL}\right)(3.8)$$

$$j_{\text{inserts}} = 1.22 \times 10^{12} \text{ ends/mL}$$

So that the production of linear ligation products can be ensured, j should be decidedly less than i. Maniatis et al. (1982) suggest that j should be 10-fold less than i such that

$$(i = 10j)$$

The j values for both the arms and the insert have now been calculated. To further ensure that j will be less than i, we will use the larger of the j values, which in this case is j_{inserts}. If we use $i = 10j$, we then have

$$i = (10)\left(1.22 \times 10^{12} \text{ ends/mL}\right) = 1.22 \times 10^{13} \text{ ends/mL}$$

Since i represents the sum of all cohesive termini in the ligation reaction ($i = i_{\text{inserts}} + i_{\text{arms}}$), the following relationship is true:

$$i = \left(2N_0M_{\text{insert}} \times 10^{-3} \text{ ends/mL}\right) + \left(2N_0M_{\text{arms}} \times 10^{-3}\text{ends/mL}\right)$$

As discussed earlier, when using a λ-derived vector, the molar concentration of annealed arms should be twice the molar concentration of the insert ($M_{\text{arms}} = 2M_{\text{insert}}$). The equation above then becomes

$$i = \left(2N_0M_{\text{insert}} \times 10^{-3} \text{ ends/mL}\right) + \left(2N_0 2M_{\text{insert}} \times 10^{-3} \text{ ends/mL}\right)$$

This equation can be rearranged and simplified to give the following relationship:

$$i = \left(2N_0M_{\text{insert}} + 2N_0 2M_{\text{insert}}\right) \times 10^{-3} \text{ ends/mL}$$

$$i = \left(2N_0M_{\text{insert}} + 4N_0M_{\text{insert}}\right) \times 10^{-3} \text{ ends/mL}$$

$$i = 6N_0M_{\text{insert}} \times 10^{-3} \text{ ends/mL}$$

The equation can be solved for M_{insert} by dividing each side of the equation by $6N_0 \times 10^{-3}$ ends/mL:

$$M_{\text{insert}} = \frac{i}{6N_0 \times 10^{-3} \text{ ends/mL}}$$

Substituting the value we have previously calculated for i (1.22×10^{13} ends/mL) and using Avogadro's number for N_0 gives the following relationship:

$$M_{\text{insert}} = \frac{1.22 \times 10^{13} \text{ ends/mL}}{6\left(6.023 \times 10^{23}\right) \times \left(1 \times 10^{-3} \text{ ends/mL}\right)}$$

$$M_{\text{insert}} = \frac{1.22 \times 10^{13} \text{ ends/mL}}{3.6 \times 10^{21} \text{ ends/mL}} = 3.4 \times 10^{-9} \ M$$

By multiplying the molarity value by the average size of the insert (20,000 bp) and the average molecular weight of a base pair (660 g/mol/bp), it is converted into a concentration expressed as µg/mL. Conversion factors are also used to convert liters to milliliters and to convert grams to micrograms. [Remember also that molarity (M) is a term for a concentration given in moles/liter; $3.4 \times 10^{-9}\ M = 3.4 \times 10^{-9}$ mol/L.]

$$x \ \mu\text{g insert/mL} = \frac{3.4\times10^{-9} \ \text{moles}}{\text{liter}} \times \frac{660 \ \text{g/mol}}{\text{bp}} \times 20{,}000 \ \text{bp} \times \frac{\text{liter}}{1000 \ \text{mL}} \times \frac{1\times10^{6} \ \mu\text{g insert}}{\text{g}}$$

$$x = \frac{44{,}880 \ \mu\text{g insert}}{1000 \ \text{mL}} = 44.9 \ \mu\text{g insert/mL}$$

The ligation reaction, therefore, should contain insert DNA at a concentration of 44.9 µg/mL. To calculate the amount of insert to place in a 150-µL reaction, the following relationship can be used. (Remember that 1 mL is equivalent to 1000 µL.)

$$\frac{x \ \mu\text{g insert}}{150 \ \mu\text{L}} = \frac{44.9 \ \mu\text{g insert}}{1000 \ \mu\text{L}}$$

$$x \ \mu\text{g insert} = \frac{(44.9 \ \mu\text{g insert})(150 \ \mu\text{L})}{1000 \ \mu\text{L}} = \frac{6735 \ \mu\text{g insert}}{1000} = 6.74 \ \mu\text{g insert}$$

Therefore, adding 6.74 µg of insert DNA to a150-µL reaction will yield a concentration of 44.9 µg insert/mL.

It was determined previously that the required molarity of lambda vector arms should be equivalent to twice the molarity of the inserts ($M_{\text{arms}} = 2M_{\text{insert}}$). The molarity of the insert fragment was calculated to be 3.4×10^{-9} M (as determined earlier). The molarity of the arms, therefore, is

$$M_{\text{arms}} = (2)(3.4\times10^{-9} \ M) = 6.8\times10^{-9} \ M$$

Therefore, the λEMBL3 arms should be at a concentration of 6.8×10^{-9} M. This value is converted to a µg arms/mL concentration, as shown in the following calculation. The 29,230-bp term represents the combined length of the two lambda arms (see Problem 9.6):

$$x \ \mu\text{g arms/mL} = \frac{6.8\times10^{-9} \ \text{moles}}{\text{liter}} \times \frac{660 \ \text{g/mol}}{\text{bp}} \times 29{,}230 \ \text{bp} \times \frac{\text{liter}}{1000 \ \text{mL}} \times \frac{1\times10^{6} \ \mu\text{g arms}}{\text{g}}$$

$$x \ \mu\text{g arms/mL} = \frac{1.3\times10^{5} \ \mu\text{g arms}}{1000 \ \text{mL}} = 130 \ \mu\text{g arms/mL}$$

Therefore, the lambda arms should be at a concentration of 130 µg/mL. The amount of λEMBL3 arms to add to a 150-µL ligation reaction can be calculated by using a relationship of ratios, as follows:

$$\frac{x\ \mu g\ \lambda EMBL3\ arms}{150\ \mu L} = \frac{130\ \mu g\ \lambda EMBL3\ arms}{1000\ \mu L}$$

$$x\ \mu g\ \lambda EMBL3\ arms = \frac{(130\ \mu g\ \lambda EMBL3\ arms)(150\ \mu L)}{1000\ \mu L} = 19.5\ \mu g\ \lambda EMBL3\ arms$$

Therefore, 19.5 μg of λEMBL3 arms should be added to 6.74 μg of insert DNA in the 150-μL ligation reaction to give optimal results.

Solution 9.6(b) The amount of λEMBL3 vector DNA to yield 19.5 μg of λEMBL3 arms can now be calculated. We will first determine the weight of a single lambda genome made up of only the two annealed arms. We will then calculate how many genomes are represented by 19.5 μg of λEMBL3 arms.

As stated in the original problem, λEMBL3 minus the stuffer fragment is 29,230 bp, and this represents one genome. The weight of one 29,230-bp genome is calculated as follows:

$$x\ \mu g/genome = 29{,}230\ bp \times \frac{660\ g/mol}{bp} \times \frac{1 \times 10^{6}\ \mu g}{g} \times \frac{1\ mol}{6.023 \times 10^{23}\ genomes}$$

$$x\ \mu g/genome = \frac{1.93 \times 10^{13}\ \mu g}{6.023 \times 10^{23}\ genomes} = 3.2 \times 10^{-11}\ \mu g/genome$$

Therefore, each 29,230-bp genome weighs 3.2×10^{-11} μg.

The number of genomes represented by 19.5 μg can now be calculated:

$$19.5\ \mu g \times \frac{1\ genome}{3.2 \times 10^{-11}\ \mu g} = 6.1 \times 10^{11}\ genomes$$

Since one genome of λEMBL3 vector will give rise to one 29,230-bp genome of lambda arms, we now need to know how many micrograms of uncut λEMBL3 DNA is equivalent to 6.1×10^{11} genomes. λEMBL3 is 42,360 bp in length. We can determine how much each λEMBL3 genome weighs by the following calculation:

$$x\ \mu g/genome = 42{,}360\ bp \times \frac{660\ g/mol}{bp} \times \frac{1 \times 10^{6}\ \mu g}{g} \times \frac{1\ mol}{6.023 \times 10^{23}\ genomes}$$

$$x\ \mu\text{g/genome} = \frac{2.8\times10^{13}\ \mu\text{g}}{6.023\times10^{23}\ \text{genomes}} = 4.6\times10^{-11}\ \mu\text{g/genome}$$

Therefore, each λEMBL3 genome weighs 4.6×10^{-11} μg. The amount of λEMBL3 DNA equivalent to 6.1×10^{11} genomes can now be calculated:

$$\frac{4.6\times10^{-11}\ \mu\text{g}}{\text{genome}}\times6.1\times10^{11}\ \text{genomes} = 28.1\ \mu\text{g}\ \lambda\text{EMBL3}$$

Therefore, digestion of 28.1 μg of λEMBL3 vector DNA will yield 19.5 μg of vector arms.

Packaging of Recombinant Lambda Genomes

Following ligation of an insert fragment to the arms of a λ vector and concatemerization of recombinant lambda genomes, the DNA must be packaged into phage protein head and tail structures, rendering them fully capable of infecting sensitive *E. coli* cells. This is accomplished by adding ligated recombinant λ DNA to a prepared extract containing the enzymatic and structural proteins necessary for complete assembly of mature virus particles. The packaging mixture is then plated with host cells that allow the formation of plaques on an agar plate (see Chapter 4). Plaques can then be analyzed for the presence of the desired clones. The measurement of the effectiveness of these reactions is called the **packaging efficiency** and can be expressed as the number of plaques (plaque-forming units, or PFU) generated from a given number of phage genomes or as the number of plaques per microgram of phage DNA.

Packaging extracts purchased commercially can have packaging efficiencies for wild-type concatemeric lambda DNA of between 1×10^{7} and 2×10^{9} PFU/μg DNA.

A method for calculating this value is demonstrated in the following problem.

Problem 9.7 The 150-μL ligation reaction described in Problem 9.6 is added to 450 μL of packaging extract. Following a period to allow for recombinant genome packaging, 0.01 mL is withdrawn and diluted to a total volume of 1 mL of lambda dilution buffer. From that tube, 0.1 mL is withdrawn and diluted to a total volume of 10 mL. One mL is withdrawn from the last dilution and is diluted further to a total volume of 10 mL dilution buffer. One-tenth mL of this last tube is plated with 0.4 mL of *E. coli* K-12 sensitive to lambda infection. Following incubation of the plate, 813 plaques are formed. What is the packaging efficiency?

Solution 9.7 The packaging reaction was diluted in the following manner:

$$\frac{0.01 \text{ mL}}{1 \text{ mL}} \times \frac{0.1 \text{ mL}}{10 \text{ mL}} \times \frac{1 \text{ mL}}{10 \text{ mL}} \times 0.1 \text{ mL plated}$$

Multiplying all numerators together and all denominators together yields the total dilution:

$$\frac{0.01 \times 0.1 \times 1 \times 0.1}{1 \times 10 \times 10} = \frac{1 \times 10^{-4} \text{ mL}}{100} = 1 \times 10^{-6} \text{ mL}$$

The concentration of viable phage in the packaging mix (given as PFU/mL) is calculated as the number of plaques counted on the assay plate divided by the total dilution:

$$\frac{813 \text{ PFU}}{1 \times 10^{-6} \text{ mL}} = 8.13 \times 10^{8} \text{ PFU/mL}$$

Therefore, the packaging mix contains 8.13×10^8 PFU/mL.

The total number of PFU in the packaging mix is calculated by multiplying the foregoing concentration by the volume of the packaging reaction [packaging mix volume = 150 μL ligation reaction + 450 μL packaging extract = 600 μL (equivalent to 0.6 mL)]:

$$\frac{8.13 \times 10^{8} \text{ PFU}}{\text{mL}} \times 0.6 \text{ mL} = 4.88 \times 10^{8} \text{ PFU total}$$

In Problem 9.6, it was determined that 6.1×10^{11} genomes could potentially be constructed from the λEMBL3 arms placed in the reaction. The percentage of these potential genomes actually packaged into viable virions can now be calculated by dividing the titer of the packaging mix by the number of potential genomes and multiplying by 100:

$$\frac{4.88 \times 10^{8}}{6.1 \times 10^{11}} \times 100 = 0.08\%$$

Therefore, 0.08% of the potential lambda genomes were packaged.

The packaging efficiency can be calculated as PFU per microgram of recombinant lambda DNA. The average recombinant phage genome should be the length of the λEMBL3 arms (29,230 bp) + the average insert (20,000 bp) = 49,230 bp. The weight of each 49,230-bp recombinant genome can be calculated by multiplying the length of the recombinant by the molecular weight of each base pair and a conversion factor that includes Avogadro's number, as follows:

$$49{,}230 \text{ bp} \times \frac{660 \text{ g/mol}}{\text{bp}} \times \frac{1 \times 10^6 \ \mu\text{g}}{\text{g}} \times \frac{1 \text{ mol}}{6.023 \times 10^{23} \text{ genomes}}$$

$$= \frac{3.25 \times 10^{13} \ \mu\text{g}}{6.023 \times 10^{23} \text{ genomes}} = 5.4 \times 10^{-11} \ \mu\text{g/genome}$$

Therefore, each 49,230-bp recombinant genome weighs 5.4×10^{-11} μg. If all 6.1×10^{11} possible genomes capable of being constructed from the λEMBL3 arms were made into 49,230-bp genomes, the total amount of recombinant genomes (in μg) would be

$$6.1 \times 10^{11} \text{ genomes} \times \frac{5.4 \times 10^{-11} \ \mu\text{g}}{\text{genome}} = 32.9 \ \mu\text{g}$$

Therefore, if all possible genomes are made and are recombinant with 20,000-bp inserts, they would weigh a total of 32.94 μg. In the packaging mix, a total of 4.88×10^8 infective viruses were made. A PFU/μg concentration is then obtained by dividing the total number of phage in the packaging mix by the total theoretical amount of recombinant DNA:

$$\frac{4.88 \times 10^8 \text{ PFU}}{32.9 \ \mu\text{g}} = 1.5 \times 10^7 \text{ PFU/}\mu\text{g}$$

Therefore, the packaging efficiency for the recombinant genomes is 1.5×10^7 PFU/μg.

Packaging extracts from commercial sources can give packaging efficiencies as high as 2×10^9 PFU/μg. These efficiencies are usually determined by using wild-type lambda DNA made into concatemers, the proper substrate for packaging. There are any number of reasons why a packaging experiment using recombinant DNA from a ligation reaction will not be as active. First, the ligation reaction may not be 100% efficient because of damaged cohesive ends or reduced ligase activity. Second, not all ligated products can be packaged; only genomes larger than 78% or smaller than 105% of the size of wild-type lambda can be packaged. This means that λEMBL3 can accommodate only fragments from about 8600 to 21,700 bp. Though ligation with the λEMBL3 arms may occur with fragments of other sizes, they will not be packaged into complete phage particles capable of infecting a cell.

Ligation Using Plasmid Vectors

Ligation of insert DNA into a bacteriophage λ-derived cloning vector requires the formation of linear concatemers of the form λ left arm:insert:λ right arm:λ left arm:insert:λ right arm:λ left arm:insert:λ right arm, etc. For this reaction to be favored, i needs to be much higher than j ($i >> j$). In contrast, ligation using a plasmid vector requires that two types of events occur. First, a linear hybrid molecule must be formed by ligation of one end of the target fragment to one complementary end of the linearized plasmid. Second, the other end of the target fragment must be ligated to the other end of the vector to create a circular recombinant plasmid. The circularization event is essential since only circular molecules transform *E. coli* efficiently. Ligation conditions must be chosen, therefore, that favor intermolecular ligation events followed by intramolecular ones.

It is suggested by Maniatis et al. (1982) that optimal results will be achieved with plasmid vectors when i is greater than j by two- to threefold. Such a ratio will favor intermolecular ligation but will still allow for circularization of the recombinant molecule. Furthermore, Maniatis et al. recommend that the concentration of the termini of the insert (i_{insert}) be approximately twice the concentration of the termini of the linearized plasmid vector ($i_{insert} = 2i_{vector}$). The following problem will illustrate the mathematics involved in calculating optimal concentrations of insert and plasmid vector to use in the construction of recombinant clones.

Problem 9.8 The plasmid cloning vector pUC19 (2686 bp) is to be cut at its single *Eco*R I site within the polylinker cloning region. This cut will linearize the vector. The insert to be cloned is a 4250-bp *Eco*R I-generated fragment. What concentration of vector and insert should be used to give an optimal yield of recombinants?

Solution 9.8 We will first calculate the values of j for both the vector and the vector + insert. These are determined by using the following equation:

$$j = j_\lambda \left(\frac{\text{MW}_\lambda}{\text{MW}} \right)^{1.5} \text{ ends/mL}$$

To use this equation for the vector, we must first calculate its molecular weight. This is accomplished by multiplying its length (in base pairs) by the conversion factor 660 g/mol/bp, as shown in the following calculation:

$$2686 \text{ bp} \times \frac{660 \text{ g/mol}}{\text{bp}} = 1.77 \times 10^6 \text{ g/mol}$$

Therefore, the pUC19 vector has a molecular weight of 1.77×10^6 g/mol.

It was determined previously that j_λ has a value of 3.22×10^{11} ends/mL and that the molecular weight of lambda (MW_λ) is equal to 3.2×10^7 g/mol. Placing these values into the earlier equation for calculating j gives the following result.

$$j = j_\lambda \left(\frac{MW_\lambda}{MW} \right)^{1.5} \text{ ends/mL}$$

$$j_{vector} = 3.22 \times 10^{11} \text{ ends/mL} \left(\frac{3.2 \times 10^7 \text{ g/mol}}{1.77 \times 10^6 \text{ g/mol}} \right)^{1.5}$$

$$j_{vector} = 3.22 \times 10^{11} \text{ ends/mL } (18.1)^{1.5}$$

$$= 3.22 \times 10^{11} \text{ ends/mL } (77) = 2.5 \times 10^{13} \text{ ends/mL}$$

Therefore, the j value for this vector is 2.5×10^{13} ends/mL. The j value for the linear ligation intermediate formed by the joining of one end of the insert fragment to one end of the linearized plasmid vector can also be calculated. The molecular weight of the ligation intermediate (vector + insert) is calculated as follows:

$$(2686 \text{ bp} + 4250 \text{ bp}) \times \frac{660 \text{ g/mol}}{\text{bp}} = 4.58 \times 10^6 \text{ g/mol}$$

Using the molecular weight of the intermediate ligation product, the j value for the intermediate can be calculated:

$$j_{insert+vector} = 3.22 \times 10^{11} \text{ ends/mL} \left(\frac{3.2 \times 10^7 \text{ g/mol}}{4.58 \times 10^6 \text{ g/mol}} \right)^{1.5}$$

$$j_{vector+insert} = 3.22 \times 10^{11} \text{ ends/mL } (6.99)^{1.5}$$

$$= 3.22 \times 10^{11} \text{ ends/mL } (18.48) = 5.95 \times 10^{12} \text{ ends/mL}$$

Therefore, the ligation intermediate has a j value of 5.95×10^{12} ends/mL. Note that the j value for the ligation intermediate ($j = 5.95 \times 10^{12}$ ends/mL) is less than that for the linearized vector alone ($j = 2.5 \times 10^{13}$ ends/mL); shorter molecules have larger j values and can circularize more efficiently than longer DNA molecules.

For optimal results, we will prepare a reaction such that i is three-fold greater than j_{vector} ($i = 3j_{vector}$). j_{vector} has been calculated to be 2.5×10^{13} ends/mL. Therefore, we want i to equal three times 2.5×10^{13} ends/mL:

$$i = 3\times\left(2.5\times10^{13}\ \text{ends/mL}\right) = 7.5\times10^{13}\ \text{ends/mL}$$

Furthermore, to favor the creation of the desired recombinant, i_{insert} should be equal to twice i_{vector} (i_{insert}:i_{vector} = 2, or $i_{insert} = 2i_{vector}$). Since i is the *total* concentration of complementary ends in the ligation reaction, $i_{insert} + i_{vector}$ must equal 7.5×10^{13} ends/mL. Since $i_{insert} = 2i_{vector}$, we have

$$2i_{vector} + i_{vector} = 7.5\times10^{13}\ \text{ends/mL}$$

which is equivalent to

$$3i_{vector} = 7.5\times10^{13}\ \text{ends/mL}$$

$$i_{vector} = \frac{7.5\times10^{13}\ \text{ends/mL}}{3} = 2.5\times10^{13}\ \text{ends/mL}$$

Therefore, i_{vector} is equivalent to 2.5×10^{13} ends/mL. Since we want i_{insert} to be two times greater than i_{vector}, we have

$$i_{insert} = 2\times\left(2.5\times10^{13}\ \text{ends/mL}\right) = 5.0\times10^{13}\ \text{ends/mL}$$

Therefore i_{insert} is equivalent to 5.0×10^{13} ends/mL. To calculate the molarity of the DNA molecules we require in the ligation reaction, we will use the equation

$$i = 2N_0M\times10^{-3}\ \text{ends/mL}$$

This expression can be rearranged to give us molarity (M), as follows:

$$M = \frac{i}{2N_0\times10^{-3}\ \text{ends/mL}}$$

For the vector, M is calculated as follows:

$$M_{\text{vector}} = \frac{2.5 \times 10^{13} \text{ ends/mL}}{2\left(6.023 \times 10^{23}\right)\left(1 \times 10^{-3} \text{ ends/mL}\right)}$$

$$M_{\text{vector}} = \frac{2.5 \times 10^{13}}{1.2 \times 10^{21}} = 2.1 \times 10^{-8} \ M$$

Therefore, the molarity of the vector in the ligation reaction should be 2.1×10^{-8} M. Converting molarity to moles/liter gives the following result:

$$2.1 \times 10^{-8} \ M = 2.1 \times 10^{-8} \text{ mol/L}$$

An amount of vector (in µg/mL) must now be calculated from the moles/liter concentration.

$$x \ \mu\text{g vector/mL} = \frac{2.1 \times 10^{-8} \text{ mol}}{\text{L}} \times \frac{660 \text{ g/mol}}{\text{bp}} \times 2686 \text{ bp} \times \frac{\text{L}}{1000 \text{ mL}} \times \frac{1 \times 10^{6} \ \mu\text{g vector}}{\text{g}}$$

$$x \ \mu\text{g vector/mL} = \frac{3.7 \times 10^{4} \ \mu\text{g vector}}{1000 \text{ mL}} = 37 \ \mu\text{g vector/mL}$$

Therefore, in the ligation reaction, the vector should be at a concentration of 37 µg/mL. The ligation reaction is going to have a final volume of 50 µL. A relationship of ratios can be used to determine how many micrograms of vector to place into the 50-µL reaction.

$$\frac{x \ \mu\text{g vector}}{50 \ \mu\text{L}} = \frac{37 \ \mu\text{g vector}}{1000 \ \mu\text{L}}$$

$$x = \frac{(50 \ \mu\text{L})(37 \ \mu\text{g vector})}{1000 \ \mu\text{L}} = 1.9 \ \mu\text{g vector}$$

Therefore, in the 50-µL ligation reaction, 1.9 µg of cut vector should be added.

It has been calculated that i_{insert} should be at a concentration of 5.0×10^{13} ends/mL. This can be converted to a molarity value by the same method used for i_{vector} using the following equation:

$$M = \frac{i}{2N_0 \times 10^{-3} \text{ ends/mL}}$$

Substituting the calculated values into this equation yields the following result:

$$M = \frac{5.0\times10^{13} \text{ ends/mL}}{2\times\left(6.023\times10^{23}\right)\times\left(1\times10^{-3} \text{ ends/mL}\right)} = \frac{5.0\times10^{13}}{1.2\times10^{21}} = 4.2\times10^{-8} \ M$$

Therefore, the insert should have a concentration of 4.2×10^{-8} *M*. This is equivalent to 4.2×10^{-8} moles of insert/L. Converting moles of insert/L to μg insert/mL is performed as follows:

$$x \ \mu\text{g insert/mL} = \frac{4.2\times10^{-8} \text{ mol}}{\text{liter}} \times \frac{660 \text{ g/mol}}{\text{bp}} \times 4250 \text{ bp} \times \frac{1 \text{ L}}{1000 \text{ mL}} \times \frac{1\times10^{6} \ \mu\text{g insert}}{\text{g}}$$

$$x \ \mu\text{g insert/mL} = \frac{1.2\times10^{5} \ \mu\text{g insert}}{1000 \text{ mL}} = 120 \ \mu\text{g insert/mL}$$

This is equivalent to 120 μg insert/1000 μL. We can determine the amount of insert to add to a 50-μL reaction by setting up a relationship of ratios, as shown here:

$$\frac{x \ \mu\text{g insert}}{50 \ \mu\text{L}} = \frac{120 \ \mu\text{g insert}}{1000 \ \mu\text{L}}$$

$$x = \frac{(50 \ \mu\text{L})\times(120 \ \mu\text{g insert})}{1000 \ \mu\text{L}} = 6 \ \mu\text{g insert}$$

Therefore, 6 μg of the 4250-bp fragment should be added to a 50-μL ligation reaction.

Transformation Efficiency

Following ligation of a DNA fragment to a plasmid vector, the recombinant molecule, in a process called **transformation**, must be introduced into a host bacterium where it can replicate. Plasmids used for cloning carry a gene for antibiotic resistance that allows for the selection of transformed cells. **Transformation efficiency** is a quantitative measure of how many cells take up plasmid. It is expressed as transformants per microgram of plasmid DNA. Its calculation is illustrated in the following problem.

Problem 9.9 A 25-μL ligation reaction contains 0.5 μg of DNA of an Ap^R (ampicillin resistance) plasmid cloning vector. 2.5 μL of the ligation reaction are diluted into sterile water to a total volume of 100 μL. Ten μL of the dilution are added to

200 μL of competent cells (cells prepared for transformation). The transformation mixture is heat shocked briefly to increase plasmid uptake, and then 1300 μL of growth medium are added. The mixture is incubated for 60 minutes to allow for phenotypic expression of the ampicillin resistance gene. Twenty μL are then spread onto a plate containing 50 μg/mL ampicillin, and the plate is incubated overnight at 37°C. The following morning, 220 colonies appear on the plate. What is the transformation efficiency?

Solution 9.9 The first step in calculating the transformation efficiency is to determine how many micrograms of plasmid DNA were in the 20-μL sample spread on the ampicillin plate. The original 25-μL ligation reaction contained 0.5 μg of plasmid DNA. We will multiply this concentration by the dilutions and the amount plated to arrive at the amount of DNA in the 20 μL used for spreading.

$$x\ \mu\text{g plasmid DNA} = \frac{0.5\ \mu\text{g plasmid DNA}}{25\ \mu\text{L}} \times \frac{2.5\ \mu\text{L}}{100\ \mu\text{L}} \times \frac{10\ \mu\text{L}}{1500\ \mu\text{L}} \times 20\ \mu\text{L}$$

$$x\ \mu\text{g plasmid DNA} = \frac{250\ \mu\text{g plasmid DNA}}{3,750,000} = 6.7 \times 10^{-5}\ \mu\text{g plasmid DNA}$$

Therefore, the 20-μL volume spread on the ampicillin plate contained 6.7×10^{-5} μg of plasmid DNA. Transformation efficiency is then calculated as the number of colonies counted divided by the amount of plasmid DNA contained within the spreading volume:

$$\text{Transformation efficiency} = \frac{220\ \text{transformants}}{6.7 \times 10^{-5}\ \mu\text{g DNA}} = \frac{3.3 \times 10^{6}\ \text{transformants}}{\mu\text{g DNA}}$$

Therefore, the transformation efficiency is 3.3×10^6 transformants/μg DNA. Competent cells prepared for transformation can be purchased from commercial sources and may exhibit transformation efficiencies greater than 1×10^9 transformants/μg DNA, as measured by using supercoiled plasmid. It should be noted that adding more than 10 ng of plasmid to an aliquot of competent cells can result in saturation and can actually decrease the transformation efficiency.

Genomic Libraries: How Many Clones Do You Need?

The haploid human genome is approximately 3×10^9 bp, contained on 23 chromosomes. The challenge in constructing a recombinant library from a genome of this size is in creating clones in large enough number so that the experimenter will be provided with some reasonable expectation that that library will contain the DNA segment carrying the particular gene of interest. How many clones need to be created to satisfy this expectation? The answer depends on two properties: (1) the size of the genome being fragmented for cloning, and (2) the average size of the cloned fragments. Clark and Carbon (1976) derived an expression, based on the Poisson distribution, that can be used to estimate the likelihood of finding a particular clone within a randomly generated recombinant library. The relationship specifies the number of independent clones, N, needed to isolate a specific DNA segment with probability P. It is given by the equation

$$N = \frac{\ln (1-P)}{\ln \left[1-\left(\frac{I}{G}\right)\right]}$$

where I is the size of the average cloned insert, in base pairs, and G is the size of the target genome, in base pairs.

Problem 9.10 In Problem 9.6, λEMBL3 is used as a vector to clone 20,000-bp fragments generated from a partial *Sau*3A1 digest of the human genome (3×10^9 bp). We wish to isolate a gene contained completely on a 20,000-bp fragment. To have a 99% chance of isolating this gene in the λEMBL3 recombinant genomic library, how many independent clones must be examined?

Solution 9.10 For this problem, $P = 0.99$ (99% probability), $I = 20{,}000$ bp, and $G = 3 \times 10^9$ bp. Placing these values into the preceding equation yields the following result:

$$N = \frac{\ln (1-0.99)}{\ln \left[1-\left(\frac{2 \times 10^4 \text{ bp}}{3 \times 10^9 \text{ bp}}\right)\right]}$$

$$N = \frac{\ln (0.01)}{\ln \left(1-6.7 \times 10^{-6}\right)} = \frac{\ln 0.01}{\ln 0.9999933} = \frac{-4.61}{-6.7 \times 10^{-6}} = 688{,}000$$

Therefore, there is a 99% chance that the gene of interest will be found in 688,000 independent λEMBL3 clones of the human genome.

This estimation, of course, is a simplification. It assumes that all clones are equally represented. For this to occur, the genome must be randomly fragmented into segments of uniform sizes, all of which can be ligated to vector with equal probability, and all of which are equally viable when introduced into host cells. When using restriction endonuclease to generate the insert fragments, this is not always possible; some sections of genomic DNA may not contain the recognition site for the particular enzyme being used for many thousands of base pairs.

cDNA Libraries: How Many Clones Are Enough?

Different tissues express different genes. There are over 500,000 mRNA molecules in most mammalian cells, representing the expression of as many as 34,000 different genes. The mRNA present in a particular tissue type can be converted to complementary DNA (cDNA) copies by the actions of reverse transcriptase (to convert mRNA to single-stranded DNA) then by DNA polymerase (to convert the single-stranded DNA to double-stranded molecules). Preparing a recombinant library from a cDNA preparation provides the experimenter with a cross section of the genes expressed in a specific tissue type.

Some genes in a cell may be expressed at a high level and be responsible for as much as 90% of the total mRNA in a cell. It can be fairly straightforward to isolate cDNA clones of the more abundant mRNA species. Up to 30% of some cell's mRNA, however, may be composed of transcripts made from genes expressed at only a very low level such that each type may be present at quantities less than 14 copies per cell. A cDNA library is not complete unless it contains enough clones that each mRNA type in the cell is represented.

The number of clones that need to be screened to obtain the recombinant of a rare mRNA at a certain probability is given by the equation

$$N = \frac{\ln\,(1-P)}{\ln\left[1-\left(\frac{n}{T}\right)\right]}$$

where

N = number of cDNA clones in the library
P = probability that each mRNA type will be represented in the library at least once (P is usually set to 0.99; a 99% chance of finding the rare mRNA represented in the cDNA library)
n = number of molecules of the rarest mRNA in a cell
T = total number of mRNA molecules in a cell

Problem 9.11 The rarest mRNA in a cell of a particular tissue type has a concentration of five molecules per cell. Each cell contains 450,000 mRNA molecules. A cDNA library is made from mRNA isolated from this tissue. How many clones will need to be screened to have a 99% probability of finding at least one recombinant containing a cDNA copy of the rarest mRNA?

Solution 9.11 For this problem, $P = 0.99$, $n = 5$, and $T = 450{,}000$. Placing these values into the preceding equation gives

$$N = \frac{\ln (1-0.99)}{\ln \left[1-\left(\frac{5}{450,000}\right)\right]}$$

$$N = \frac{\ln (0.01)}{\ln \left(1-1.1\times10^{-5}\right)} = \frac{-4.61}{-1.1\times10^{-5}} = 420,000$$

Therefore, in a cDNA library of 420,000 independent clones, one clone should represent the rarest mRNA in that particular cell type.

Expression Libraries

A number of cloning vectors have been designed with genetic elements that allow for the expression of foreign genes. These elements usually include the signals for transcription and translation initiation. For many expression vectors, the cloning site is downstream from an ATG start signal. For example, a truncated version of the beta-galactosidase gene from *E. coli* is often used as a source for translation signals. Insertion of a target fragment in the proper orientation and translation frame results in the production of a fusion protein between the truncated gene on the vector and the foreign inserted gene.

Expression libraries can be made directly from bacterial genomic DNA since microbial genes characteristically lack introns. Mammalian gene expression libraries, however, are best constructed from cDNA, which, since it is made from processed mRNA, is free of intron sequences.

No matter what the source of target fragment, expression relies on the positioning of the insert in the correct orientation and the correct reading frame with the vector's expression elements. The probability of obtaining a recombinant clone with the insert fragment positioned correctly in the expression vector is given by the equation

$$P = \frac{\text{coding sequence size (in kb)}}{\text{genome size (in kb)} \times 6}$$

The denominator in this equation contains the factor 6 since there are six possible positions the target fragment can insert in relation to the vector's open reading frame. The target fragment can be inserted in either the sense or antisense orientation and in any one of three possible reading frames. One of the six possible positions should be productive.

The proportion of positive recombinants expressing the gene on the target fragment is given by the expression $1/P$.

This equation provides only an estimate. For randomly sheared genomic fragments, optimal results will be achieved if the cloning fragments are roughly one-half the size of the actual gene's coding sequence, because smaller fragments are less likely to carry upstream termination signals. For the cloning of fragments from large genomes containing many introns, such as those from humans, a large number of clones must be examined.

Problem 9.12 An *E. coli* gene has a coding sequence spanning 2000 bp. It is to be recovered from a shotgun cloning experiment in which sheared fragments of genomic DNA are inserted into an expression vector. What proportion of recombinants would be expected to yield a fusion protein with the vector?

Solution 9.12 Two thousand bp is equivalent to 2 kb:

$$2000 \text{ bp} \times \frac{1 \text{ kb}}{1000 \text{ bp}} = 2 \text{ kb}$$

The *E. coli* genome is 4,640,000 bp in length. This is equivalent to 4640 kb. Placing these values into the equation gives

$$P = \frac{2 \text{ kb}}{(4640 \text{ kb}) \times 6} = \frac{2}{2.8 \times 10^{4}} = 7.1 \times 10^{-5}$$

Taking the reciprocal of this value gives the following result:

$$\frac{1}{P} = \frac{1}{7.1 \times 10^{-5}} = 1.4 \times 10^{4}$$

Therefore, 1 out of approximately 14,000 clones will have the desired insert in the proper orientation to produce a fusion protein.

Screening Recombinant Libraries by Hybridization to DNA Probes

Either plaques on a bacterial lawn or colonies spread on an agar plate can be screened for the presence of clones carrying the desired target fragment by hybridization to a labeled DNA probe. In this technique, plaques or colonies are transferred to a nitrocellulose or nylon membrane. Once the DNA from the recombinant clones is fixed to the membrane (by baking in an oven or by UV cross-linking), the membrane is placed in a hybridization solution containing a radioactively labeled DNA probe designed to anneal specifically to sequence of the desired target fragment. Following incubation with probe, the membrane is washed to remove any probe bound nonspecifically to the membrane and X-ray film is placed over the membrane to identify those clones annealing to probe.

Annealing of probe to complementary sequence follows association/dissociation kinetics and is influenced by probe length, G/C content, temperature, salt concentration, and the amount of formamide in the hybridization solution. These factors affect annealing in the following way. Lower G/C content, increased probe length, decreased temperature, and higher salt concentrations all favor probe association to target. Formamide is used in a hybridization solution as a solvent. It acts to discourage intrastrand annealing.

The stability of the interaction between an annealing probe and the target sequence as influenced by these variables is best measured by their affect on the probe's melting temperature (T_m), the temperature at which half of the probe has dissociated from its complementary sequence. Estimation of a probe's melting temperature is important to optimizing the hybridization reaction. Davis et al. (1986) recommend the temperature at which hybridization is performed (T_i) to be 15°C below T_m:

$$T_i = T_m - 15°\text{C}$$

T_m is calculated using the equation

$$T_m = 16.6 \log[M] + 0.41[P_{GC}] + 81.5 - P_m - B/L - 0.65[P_f]$$

where

M = molar concentration of Na^+, to a maximum of 0.5 M (1X SSC contains 0.165 M Na^+)

P_{GC} = percent of G and C bases in the oligonucleotide probe (between 30% and 70%)

P_m = percent of mismatched bases, if known (each percent of mismatch Will lower T_m by approximately 1°C)

P_f = percent formamide

B = 675 (for synthetic probes up to 100 nucleotides in length)

L = probe length in nucleotides

For probes longer that 100 bases, the following formula can be used to determine T_m:

$$T_m = 81.5°\text{C} + 16.6(\log[\text{Na}^+]) + 0.41(\%\ \text{GC}) - 0.63(\%\ \text{formamide}) - 600/L$$

For these longer probes, hybridization is performed at 10°C to 15°C below T_m.

Problem 9.13 A labeled probe has the following sequence:

5′– GAGGCTTACGCGCATTGCCGCGATTTGCCCATCGCAAGTACGCAATTAGCAC –3′

It will be used to detect a positive clone from plates of λEMBL3 recombinant plaques transferred to a nitrocellulose membrane filter. Hybridization will take place in 1X SSC (0.15 *M* NaCl, 0.015 *M* Na_3Citrate • $2H_2O$) containing 50% formamide. At what temperature should hybridization be performed? (Assume no mismatches between the probe and the target sequence.)

Solution 9.13 The probe is 52 nucleotides in length. It contains a total of 29 G and C residues. Its P_{GC} value, therefore, is

$$P_{GC} = \frac{29 \text{ nucleotides}}{52 \text{ nucleotides}} \times 100 = 56\%$$

Both sodium chloride (0.15 *M*) and sodium citrate (0.015 *M*) in the hybridization buffer contribute to the overall sodium ion concentration. The total sodium ion concentration is

$$0.15\ M + 0.015\ M = 0.165\ M$$

Using these values, the equation for T_m of the 52-mer probe is calculated as follows:

$$T_m = 16.6 \log[0.165] + 0.41\ [56] + 81.5° - 0 - 675/52 - 0.65[50]$$

$$T_m = 16.6[-0.783] + 22.96 + 81.5° - 12.98 - 32.5$$

$$T_m = 46°\text{C}$$

The hybridization temperature (T_i) is calculated as follows:

$$T_i = 46°\text{C} - 15°\text{C} = 31°\text{C}$$

Therefore, the hybridization should be performed at 31°C.

Oligonucleotide Probes

Probes used for hybridization to DNA bound to nitrocellulose or nylon membranes is either synthetic (an oligonucleotide prepared on an instrument) or a nucleic acid amplified by cloning or PCR from a biological source. If the exact gene sequence of the target fragment is known, a synthetic oligonucleotide can be made having perfect complementarity.

The number of sites on a genome to which an oligonucleotide probe will anneal depends on the genome's complexity and the number of mismatches between the probe and the target. A mismatch is a position within the probe/target hybrid occupied by noncomplementary bases. Complexity (C) is the size of the genome (in base pairs) represented by unique (single copy) sequence and one copy of each repetitive sequence region. The genomes of most bacteria and single-celled organisms do not have many repetitive sequences. The complexity of such organisms, therefore, is similar to genome size. Mammalian genomes, however, from mice to men, contain repetitive sequences in abundance. Although most mammalian genomes are roughly 3×10^9 bp in size, Laird (1971) has estimated their complexity to be approximately equivalent to 1.8×10^9 bp.

The number of positions on a genome having a certain complexity to which an oligonucleotide probe will hybridize is given by the equation

$$P_0 = \left(\frac{1}{4}\right)^L \times 2C$$

where P_0 is the number of independent perfect matches, L is the length of the oligonucleotide probe, and C is the target genome's complexity (this value is multiplied by 2 to represent the two complementary strands of DNA, either of which could potentially hybridize to the probe) (Maniatis et al., 1982).

Problem 9.14 An oligonucleotide probe is 12 bases in length. To how many positions on human genomic DNA could this probe be expected to hybridize?

Solution 9.14 Using the preceding equation, L is equal to 12 and C is equivalent to 1.8×10^9. Placing these values into the equation gives the following result:

$$P_0 = \left(\frac{1}{4}\right)^{12} \times 2\left(1.8\times10^9\right) = \left(6\times10^{-8}\right)\times\left(3.6\times10^9\right) = 216$$

Therefore, by chance, an oligonucleotide12 bases long should hybridize to 216 places on the human genome.

Problem 9.15 What is the minimum length an oligonucleotide should be to hybridize to only one site on the human genome?

Solution 9.15 In Problem 9.14, it was assumed that the term $2C$ for the human genome is equivalent to 3.6×10^9. We want P_0 to equal 1. Our equation then becomes

$$1 = \left(\frac{1}{4}\right)^L \times 3.6\times10^9$$

$$\frac{1}{3.6 \times 10^9} = (0.25)^L$$

Convert the fraction 1/4 into a decimal and divide each side of the equation by 3.6×10^9.

$$2.8 \times 10^{-10} = (0.25)^L$$

$$\log 2.8 \times 10^{-10} = \log 0.25^L$$

Take the common logarithm of both sides of the equation.

$$\log 2.8 \times 10^{-10} = L(\log 0.25)$$

Using the power rule for logarithms (for any positive numbers M and a (where a is not equal to 1), and any real number p, the logarithm of a power of M is the exponent times the logarithm of M: $\log_a M^p = p \log_a M$).

$$\frac{\log 2.8 \times 10^{-10}}{\log 0.25} = L$$

Divide each side of the equation by log 0.25.

$$\frac{-9.6}{-0.6} = 16$$

Take the log values.

Therefore, an oligonucleotide 16 bases in length would be expected to hybridize to the human genome at only one position.

Hybridization Conditions

Determination of an oligonucleotide probe's melting temperature (T_m) can help the experimenter decide on a temperature at which to perform a hybridization step. Hybridization is typically performed at 2–5°C below the oligonucleotide's T_m. However, the length of time that the probe is allowed to anneal to target sequence is another aspect of the reaction that must be considered. For an oligonucleotide probe, the length of time allowed for the hybridization reaction to achieve half-completion is given by the equation described by Wallace and Miyada (1987):

$$t_{1/2} = \frac{\ln 2}{kC}$$

where k is a first-order rate constant and C is the molar concentration of the oligonucleotide probe (in moles of nucleotide per liter). This equation should be used only as an approximation. The actual rate of hybridization can be three to four times slower than the calculated rate (Wallace et al., 1979).

The rate constant, k, represents the rate of hybridization of an oligonucleotide probe to an immobilized target nucleic acid in 1 M sodium ion. It is given by the following equation:

$$k = \frac{3 \times 10^5 L^{0.5} \text{ L/mol/s}}{N}$$

where k is calculated in liters/mole of nucleotide per second, L is the length of the oligonucleotide probe in nucleotides, and N is the probe's complexity.

Complexity, as it relates to an oligonucleotide, is calculated as the number of different possible oligonucleotides in a mixture. The oligonucleotide 5′– GGACCTATAGCCGTTGCG –3′, an 18-mer, since it has only one defined sequence, has a complexity of 1. When oligonucleotides are constructed by reverse translation of a protein sequence, as might be the case when probing for a gene for which there exists only protein sequence information, a degenerate oligonucleotide might be synthesized having several possible different sequences at the wobble base position. For example, reverse translation of the following protein sequence would require the synthesis of an oligonucleotide with three possible sequences:

```
MetTrpMetIleTrpTrp
ATGTGGATGATATGGTGG
           C
           T
```

Methionine (Met) and tryptophan (Trp) are each encoded by a single triplet. Isoleucine (Ile), however, can be encoded for by three possible triplets, ATA, ATC, and ATT. The oligonucleotide is synthesized as a mixed probe containing either an A, a C, or a T at the 12th position. Since there are three possible sequences for the oligonucleotide, it has a complexity of 3.

When designing a probe based on protein sequence, the experimenter should choose an area to reverse translate that has the least amount of codon redundancy. Nevertheless, since most amino acids are encoded for by two or more triplets (see Table 9.1), most mixed probes will contain a number of redundancies. The total number of different oligonucleotides in a mixture, its complexity, is calculated by multiplying the number of possible nucleotides at all positions. For example, here is a protein sequence reverse translated using all possible codons:

```
ArgProLysPheTrpIleCysAla
AGACCAAAATTCTGGATATGCGCA
C C  C  G  T     C  T  C
  G  G           T     G
  T  T                 T
```

To determine how many different sequences are possible in such a mixture, the number of possible nucleotides at each position are multiplied together. In this example, the total number of different sequences is calculated as

$$2 \times 1 \times 4 \times 1 \times 1 \times 4 \times 1 \times 1 \times 2 \times 1 \times 1 \times 2 \times 1 \times 1 \times 1 \times 1 \times 1 \times 3 \times 1 \times 1 \times 2 \times 1 \times 1 \times 4 = 3072$$

Therefore, there are 3072 possible different oligonucleotides in this mixture.

Amino Acid	Abbreviation	Codons
Alanine	Ala	GCA GCC GCG GCT
Arginine	Arg	AGA AGG CGA CGC CGG CGT
Asparagine	Asn	AAC AAT
Aspartic acid	Asp	GAC GAT
Cysteine	Cys	TGC TGT
Glutamine	Gln	CAA CAG
Glutamic acid	Glu	GAA GAG
Glycine	Gly	CGA GGC GGG GGT
Histidine	His	CAC CAT
Isoleucine	Ile	ATA ATC ATT
Leucine	Leu	CTA CTC CTG CTT TTA TTG
Lysine	Lys	AAA AAG
Methionine	Met	ATG
Phenylalanine	Phe	TTC TTT
Proline	Pro	CCA CCC CCG CCT
Serine	Ser	TCA TCC TCG TCT AGC AGT
Threonine	Thr	ACA ACC ACG ACT
Tryptophan	Trp	TGG
Tyrosine	Tyr	TAC TAT
Valine	Val	GTA GTC GTG GTT
STOP		TAA TAG TGA

Table 9.1 The genetic code. Codons are shown as DNA sequence rather than RNA sequence (T's are used instead of U's) to facilitate their conversion to synthetic oligonucleotide probes.

Problem 9.16 An oligonucleotide probe 21 nucleotides in length is used for hybridization to a colony filter containing recombinant clones. The oligonucleotide probe is a perfect match to the target gene to be identified. The hybridization will be performed in 1 *M* sodium chloride at 2°C below the probe's T_m. The hybridization solution contains oligonucleotide probe at a concentration of 0.02 μg/mL. **(a)** When will hybridization be half-complete? **(b)** How long should hybridization be allowed to proceed?

Solution 9.16(a) We will first calculate the value for *k*, the rate constant, using the following expression:

$$k = \frac{3 \times 10^5 L^{0.5} \ \text{L/mol/s}}{N}$$

For this problem, *L*, the oligonucleotide length is 21. Since the probe is an exact match, *N*, the complexity of the oligonucleotide, is equal to 1. Placing these values into the equation for *k* gives

$$k = \frac{(3 \times 10^5)(21)^{0.5} \ \text{L/mol/s}}{1} = (3 \times 10^5)(4.58) = 1.37 \times 10^6 \ \text{L/mol/s}$$

So that this *k* value can be used in the equation for calculating the half-complete hybridization reaction, a value for *C*, the molar concentration of the oligonucleotide probe, must also be calculated. In this problem, the probe has a concentration of 0.02 μg/mL. To convert this to a molar concentration (moles/liter), we must first determine the molecular weight (grams/mole) of the probe. Since the oligonucleotide probe is 21 nucleotides (nt) in length, we can determine its molecular weight by multiplying the length in nucleotides by the molecular weight of a single nucleotide:

$$21 \ \text{nt} \times \frac{330 \ \text{g/mol}}{1 \ \text{nt}} = 6930 \ \text{g/mol}$$

Therefore, the molecular weight of a 21-mer is approximately 6930 g/mol. This value can then be used to calculate the molarity of the probe in the hybridization solution by the use of several conversion factors, as shown in the following equation.

$$\frac{0.02 \ \mu\text{g}}{\text{mL}} \times \frac{1000 \ \text{mL}}{\text{L}} \times \frac{1 \ \text{mol}}{6930 \ \text{g}} \times \frac{1 \ \text{g}}{1 \times 10^6 \ \mu\text{g}} = C \ \text{mol/L}$$

$$\frac{20 \text{ mol}}{6.93 \times 10^9 \text{ L}} = C \text{ mol/L}$$

$$C = 2.89 \times 10^{-9} \text{ mol/L}$$

The values calculated for k and C can now be used in the expression for half-complete hybridization.

$$t_{1/2} = \frac{\ln 2}{(1.37 \times 10^6 \text{ L/mol/s})(2.89 \times 10^{-9} \text{ mol/L})}$$

$$t_{1/2} = \frac{0.693}{\dfrac{3.96 \times 10^{-3}}{1 \text{ s}}} = 175 \text{ s}$$

Therefore, hybridization is half-complete in 175 seconds.

Solution 9.16(b) Since the actual rate of hybridization can be approximately four times slower than the calculated value (Wallace et al., 1979), the value calculated in Solution 9.16(a) can be multiplied by 4:

$$175 \text{ seconds} \times 4 = 700 \text{ seconds}$$

Converting this to minutes gives the following result:

$$700 \text{ seconds} \times \frac{1 \text{ minute}}{60 \text{ seconds}} = 11.7 \text{ minutes}$$

Therefore, the hybridization reaction described in this problem is half-complete in about 11 minutes, 42 seconds.

Problem 9.17 A small amount of a novel protein is purified and sequenced. A region of the protein is chosen from which an oligonucleotide probe is to be designed. The sequence of that region is as follows:

TrpTyrMetHisGlnLysPheAsnTrp

The mixed probe synthesized from this protein sequence will be added at a concentration of 0.02 μg/mL to a hybridization solution containing 1 *M* sodium chloride for the purpose of identifying a recombinant clone within a library. Hybridization will be performed at a temperature several degrees below the average T_m of the mixed probe. When will the hybridization be half-complete?

Solution 9.17 The first step in solving this problem is to determine the complexity, N, of the reverse-translated mixed probe that would be extrapolated from the protein sequence. Using Table 9.1, the protein fragment is reverse translated as follows:

```
TrpTyrMetHisGlnLysPheAsnTrp
TGGTACATGCACCAAAAATTCAACTGG
    T     T  G  G  T  T
```

The number of possible oligonucleotides in the mixed oligo pool is calculated by multiplying all possible codons at each amino acid position (we will include here only those positions having greater than 1 possibility). This gives

$$2 \times 2 \times 2 \times 2 \times 2 \times 2 = 64$$

Therefore, the mixed oligonucleotide synthesized as a probe has a complexity, N, equal to 64. The oligonucleotide probe, no matter which base is at any position, has a length (L) of 27 nucleotides.

The value for k, the hybridization rate, for this oligonucleotide is calculated as follows:

$$k = \frac{3 \times 10^5 L^{0.5}\ \text{L/mol/s}}{N}$$

Substituting the given values for N and L into the equation yields the following result:

$$k = \frac{(3 \times 10^5)(27)^{0.5}\ \text{L/mol/s}}{64}$$

$$k = \frac{(3 \times 10^5)(5.2)\ \text{L/mol/s}}{64} = \frac{1.6 \times 10^6\ \text{L/mol/s}}{64} = 2.5 \times 10^4\ \text{L/mol/s}$$

The probe has a molecular weight calculated as follows:

$$27\ \text{nucleotides} \times \frac{330\ \text{g/mol}}{\text{nucleotide}} = 8910\ \text{g/mol}$$

The molarity of the probe, when at a concentration of 0.02 μg/mL in the hybridization solution, is calculated as follows:

$$\frac{0.02\ \mu\text{g probe}}{\text{mL}} \times \frac{1000\ \text{mL}}{\text{L}} \times \frac{1\ \text{mol}}{8910\ \text{g}} \times \frac{1\ \text{g}}{1 \times 10^{6}\ \mu\text{g}} = C\ \text{mol/L}$$

$$\frac{20\ \text{mol probe}}{8.91 \times 10^{9}\ \text{L}} = C\ \text{mol/L}$$

$$C = 2.24 \times 10^{-9}\ \text{mol/L}$$

Placing the calculated values for k and C into the equation for half-complete hybridization gives the following result.

$$t_{1/2} = \frac{\ln 2}{\left(2.5 \times 10^{4}\ \text{L/mol/s}\right)\left(2.24 \times 10^{-9}\ \text{mol/L}\right)}$$

$$t_{1/2} = \frac{0.693}{\frac{5.6 \times 10^{-5}}{\text{s}}} = 12{,}375\ \text{s}$$

Since, as discussed previously, the actual half-complete hybridization reaction can be as much as four times longer than the calculated value, we will allow for maximum chances for hybridization between probe and target to proceed by multiplying our value by 4:

$$12{,}375\ \text{seconds} \times 4 = 49{,}500\ \text{seconds}$$

Converting this answer into hours gives the following result:

$$49{,}500\ \text{seconds} \times \frac{1\ \text{minute}}{60\ \text{seconds}} \times \frac{1\ \text{hour}}{60\ \text{minutes}} = 13.75\ \text{hours}$$

Therefore, using this mixed probe and the appropriate hybridization temperature, the hybridization reaction should be half-complete in 13.75 hours.

Hybridization Using Double-Stranded DNA Probes

Hybridization can also be performed using a double-stranded DNA probe prepared by nick translation. If you use a hybridization temperature of 68°C in aqueous solution or 42°C in 50% formamide, Maniatis et al. (1982) recommend the use of the following equation to estimate the amount of time to achieve half-complete hybridization:

$$t_{1/2} = \frac{1}{X} \times \frac{Y}{5} \times \frac{Z}{10} \times 2$$

where

X = amount of probe added to the hybridization reaction (in μg)
Y = probe complexity, which for most probes is proportional to the length of the probe, in kilobases (kb)
Z = volume of the hybridization reaction (in mL)

Nearly complete hybridization is achieved after three times the $t_{1/2}$ period.

Problem 9.18 Filters prepared for plaque hybridization are placed in a plastic bag with 15 mL of an aqueous hybridization buffer and 0.5 μg of a nick-translated probe 7500 bp in length. Hybridization is allowed to proceed at 68°C. How long should the hybridization reaction continue for nearly complete hybridization of probe to target?

Solution 9.18 Using the preceding equation, X is 0.5 μg, Y is 7.5 kb (7500 bp = 7.5 kb), and Z is 15 mL. Placing these values into the equation for half-complete hybridization gives

$$t_{1/2} = \frac{1}{0.5} \times \frac{7.5}{5} \times \frac{15}{10} \times 2 = \frac{225}{25} = 9$$

Therefore, the hybridization reaction is half-complete in 9 hours. The reaction can be allowed to proceed for three times as long (27 hours) to ensure nearly complete hybridization.

Sizing DNA Fragments by Gel Electrophoresis

Once a candidate recombinant clone has been identified by either colony hybridization to an allele specific probe or by expression of a desired protein, it should be adequately characterized to ensure it contains the expected genetic material. One of the most straightforward and essential characterizations of a nucleic acid is the determination of its length. Fragment sizing is usually performed after a PCR amplification or in the screening of recombinant clone inserts that have been excised by restriction digestion.

Fragment sizing is accomplished by electrophoresing the DNA fragments in question on an agarose or acrylamide gel in which one lane is used for the separation of DNA fragments of known size (molecular weight or size markers). Following gel staining, the migration distance of each band of the size marker is measured. These values are plotted on a semilog graph. The rate at which a DNA fragment travels during electrophoresis is inversely proportional to the $\log_{10}$ of its length in base pairs. The graph generated for the size markers should contain a region represented by a straight line where this relationship is most pronounced. Larger DNA fragments show this relationship to a lesser degree, depending on the concentration of the gel matrix. Agarose is an effective gel matrix to use for separation of DNA fragments ranging in length between 50 and 30,000 base pairs. The experimenter, however, should use a percentage of agarose suitable for the fragments to be resolved. Low-molecular-weight fragments should be run on higher-percentage gels (1% to 4% agarose). High-molecular-weight fragments are best resolved on low-percentage gels (0.3% to 1% agarose). The best resolution of low-molecular-weight fragments can be achieved on a polyacrylamide gel. No matter what type of gel matrix is used, the method for determining fragment size is the same.

Figure 9.1 shows a depiction of an agarose gel run to determine the sizes of cDNA fragments inserted into a plasmid vector. On this gel, a 100-bp ladder is run in the first lane as a size marker.

In Problem 9.19, the fragment size of a cDNA insert will be determined by two methods. In the first approach, a semilog graph and a standard curve will be used to extrapolate the fragment size. In the second approach, linear regression performed on Microsoft Excel software will be used to derive a best-fit line from a standard curve and the line's equation will be used to calculate the fragment's size.

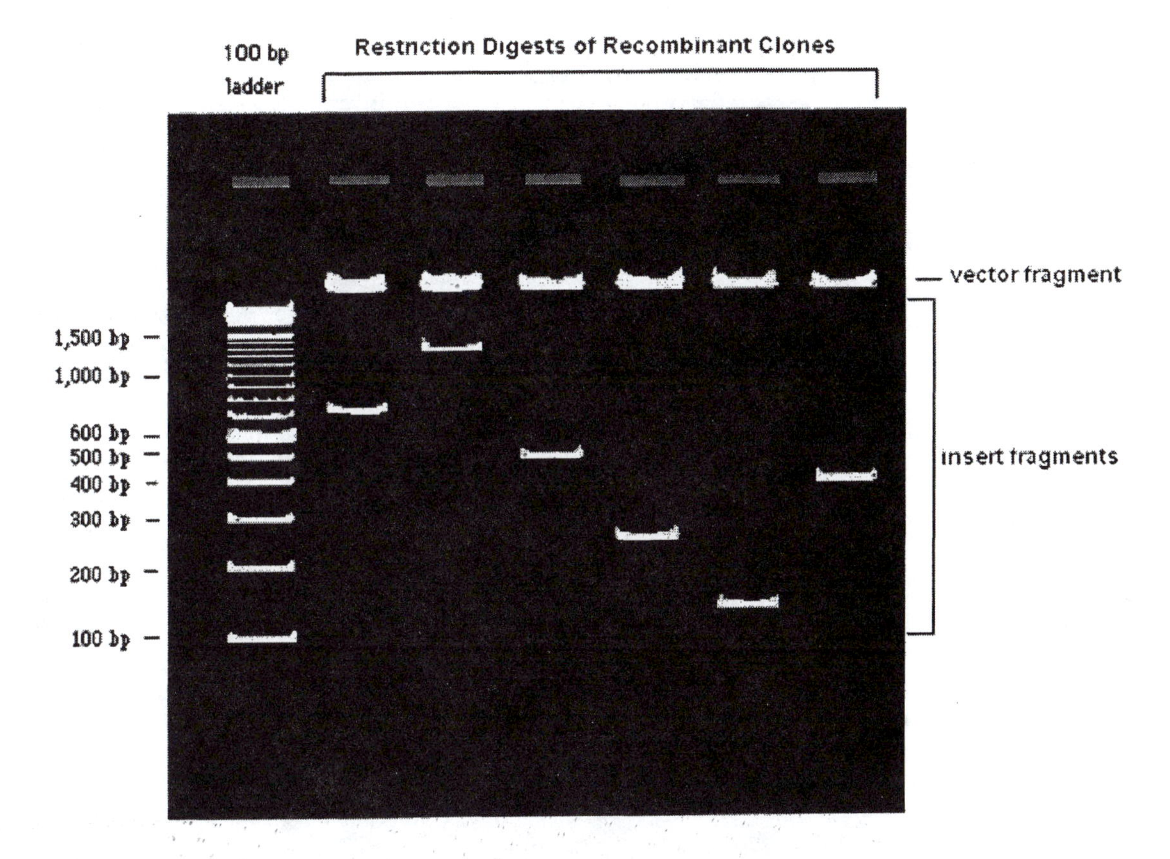

Figure 9.1 Representation of an agarose gel used to size cDNA insert fragments. DNA of cDNA clones is digested with restriction endonucleases at unique sites on both sides of the insert. The digests are electrophoresed to separate the vector fragment from the insert fragment. The 100-bp ladder loaded in lane 1 is a size marker in which each fragment differs in size by 100 base pairs from the band on either side of it. For this size marker, the 600-bp and 1000-bp bands stain more intensely than the others.

To determine fragment sizes on an agarose gel, a ruler is placed on the gel photograph along the size markers such that the zero-centimeter mark is aligned with the bottom of the well (Figure 9.2).

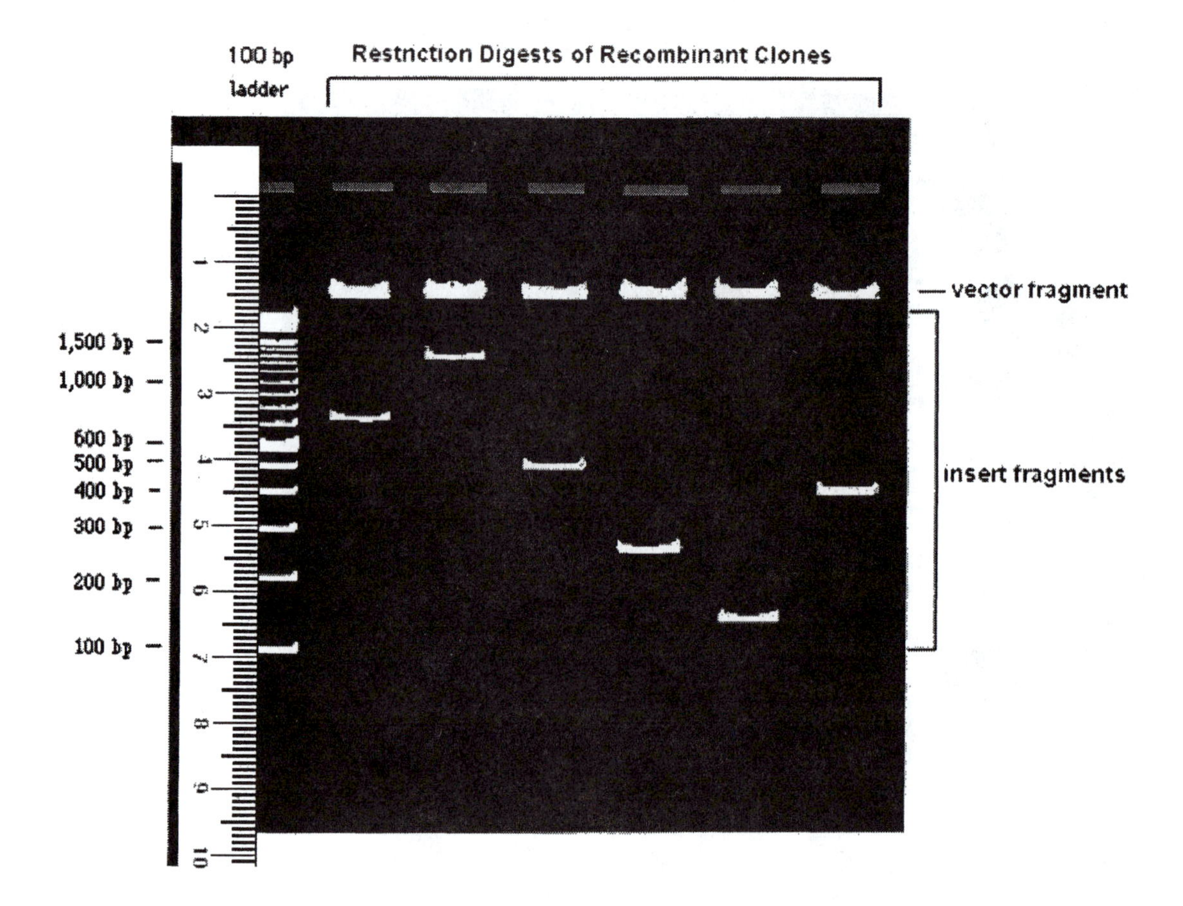

Figure 9.2 Measuring the distance of each band in the 100-bp ladder size marker from the well.

The distance from the bottom of the well to the bottom of each band is then measured and recorded (Table 9.2).

Size Marker Band (bp)	Distance from Well (cm)
1500	2.3
1400	2.35
1300	2.45
1200	2.55
1100	2.7
1000	2.85
900	3.0
800	3.2
700	3.5
600	3.9
500	4.15
400	4.55
300	5.1
200	5.85
100	7.0

Table 9.2 Migration distances for fragments of the100-bp ladder, as shown in Figure 9.2.

The distance from the well to each band is then plotted on a semilog graph and a straight line is drawn on the plot connecting as many points as possible (Figure 9.3).

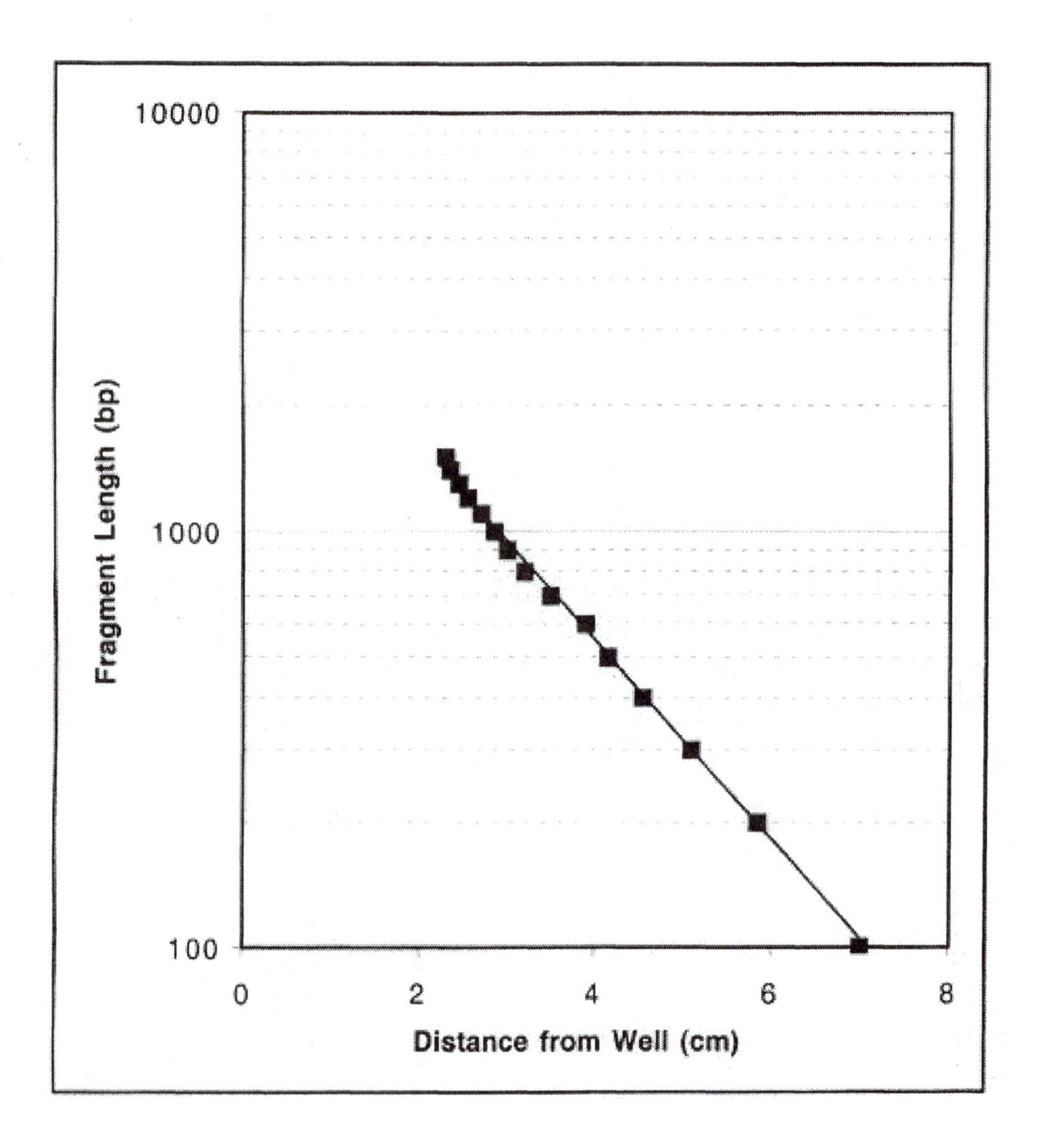

Figure 9.3 Plot of distance of each band of the 100-bp ladder from the well on a semilog graph.

Problem 9.19 What is the size of the insert shown in lane 4 of the gel in Figure 9.1? Determine the fragment's size (**a**) using a semilog plot and (**b**) by linear regression using Microsoft Excel.

Solution 9.19(a) Place a ruler on lane 4 of the photograph and measure the distance from the well to the bottom of the band representing the insert fragment (Figure 9.4).

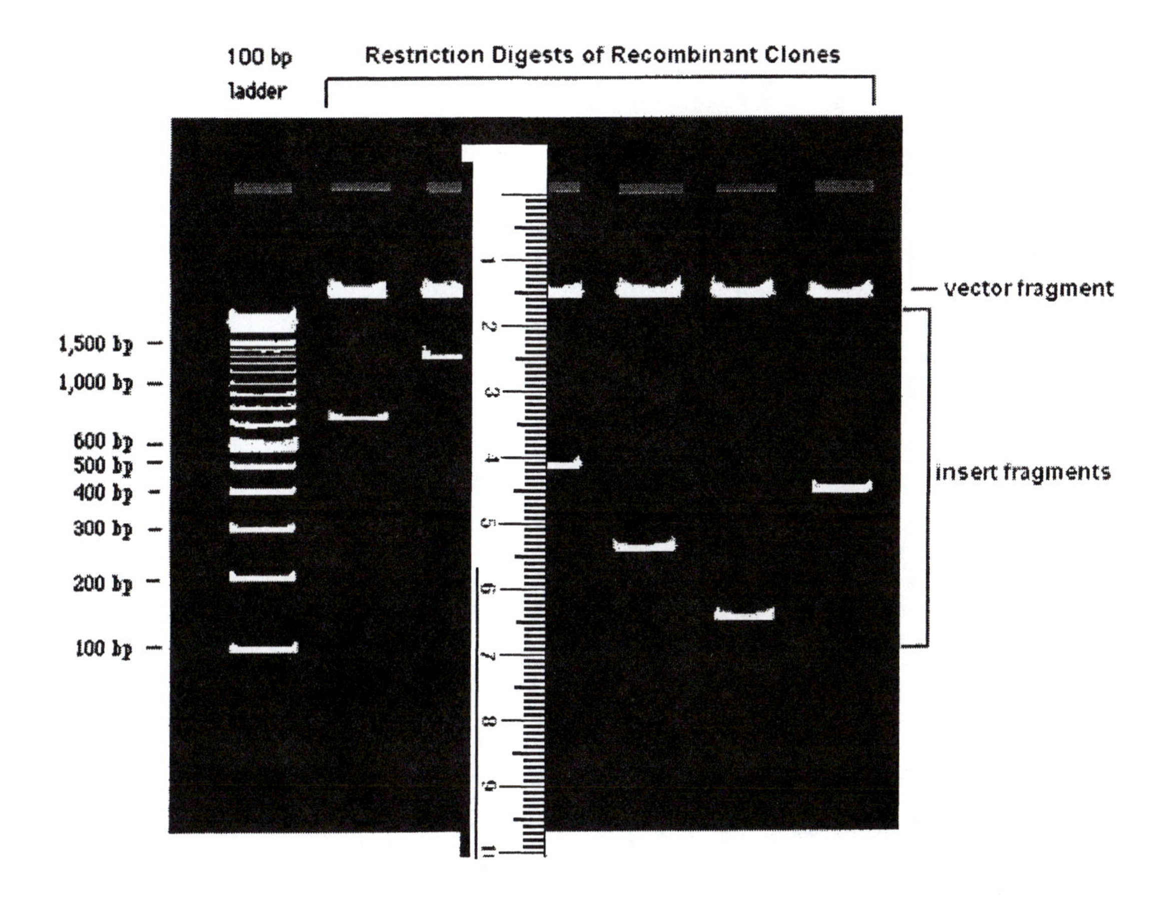

Figure 9.4 Measuring the distance of the insert band from the well for lane 4.

The band measures 4.2 cm from the well. On the semilog plot of the standard curve (Figure 9.5), 4.2 cm on the *x* axis corresponds to 470 bp on the *y* axis.

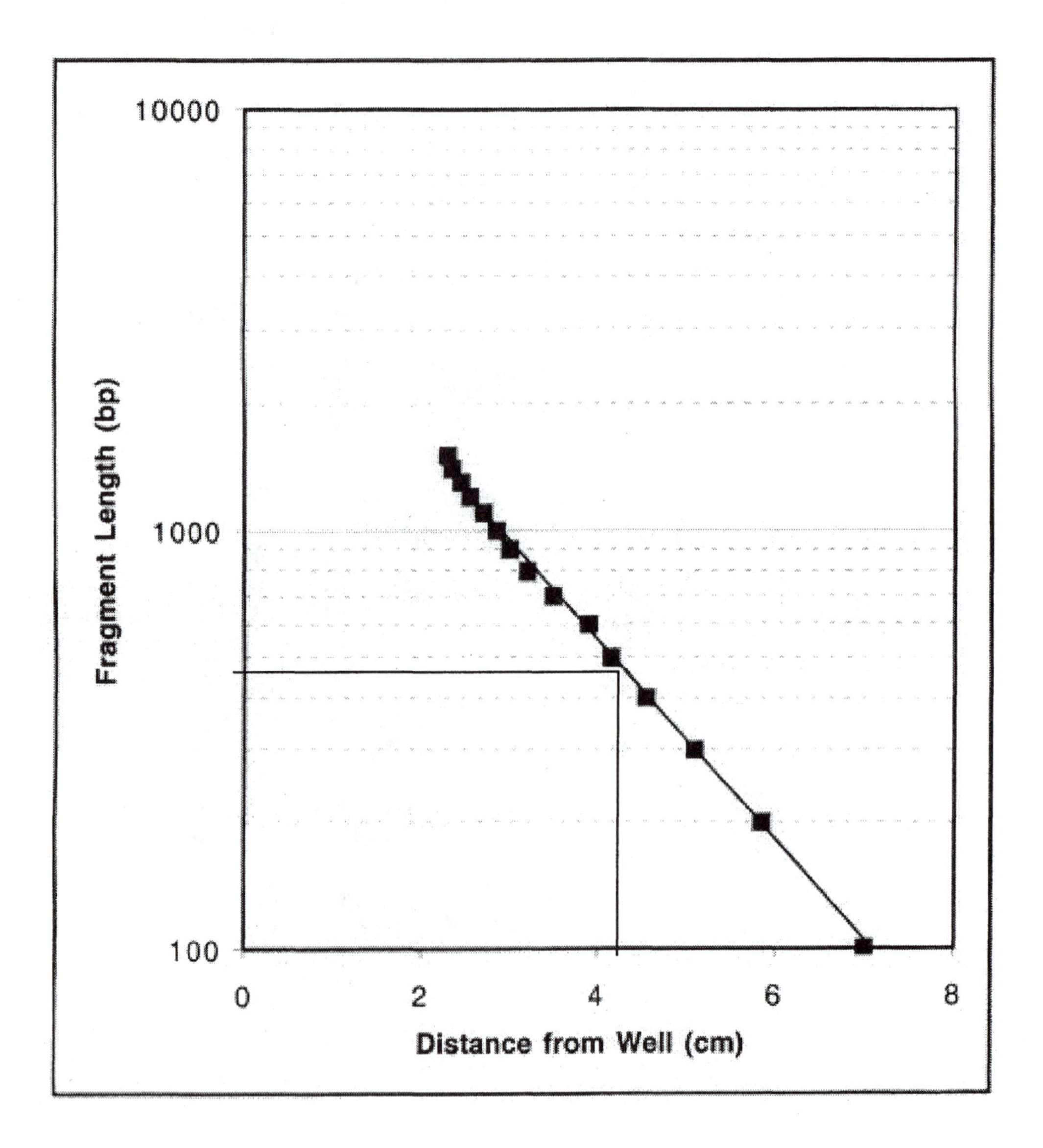

Figure 9.5 Extrapolating the size of the DNA fragment in Lane 4 of Figure 9.1. A vertical line is drawn from 4.2 cm on the *x* axis to the curve, and then a horizontal line is drawn to the *y* axis to determine the fragment's size in base pairs.

Therefore, the insert fragment in lane 4 is approximately 470 bp in length.

Solution 9.19(b) The Microsoft Office software package is one of the most popular computer applications for home and business use. Included in this software package is the Excel program. Although designed for inventory maintenance and for creating spreadsheets to track financial transactions, it is equipped with math functions that the molecular biologist might find useful. In particular, it can perform linear regression, which can be used to calculate fragment sizes.

Linear regression is a mathematical technique for deriving a line of best fit, as well as the equation that describes that line, for points on a plot showing the relationship between two variables. A line of best fit, or **regression line**, is a straight line drawn through points on a plot such that the line passes through (or as close to) as many points as possible. Linear regression, therefore, is used to determine if a direct and linear relationship exists between two variables. In this case, the two variables are fragment size and distance migrated from the well during electrophoresis.

DNA fragments migrate through an agarose gel at a rate that is inversely proportional to the logarithm (base 10) of their length. This means, in practical terms, that a straight-line relationship exists between the distance a fragment has migrated on a gel and the log of the fragment's size (in base pairs). This can be seen graphically if a plot of migration distance versus base pair size is drawn on a semilog graph [as shown in Solution 9.19(a)] or if migration distance versus log of fragment size is graphed on a standard plot. By either technique, a straight line describes the relationship between these two variables.

A straight line is defined by the general equation

$$y = mx + b$$

For use of this equation to calculate fragment sizes by linear regression

y = log of the fragment size [plotted on the vertical (y) axis]
m = slope of the regression line
x = distance (in cm) from the well [plotted on the horizontal (x) axis]
b = point where the regression line intercepts the y axis

A measure of how closely the points on a plot fall along a regression line is given by the correlation coefficient (represented by the symbol R^2). R^2 will have a positive value if the slope of the regression line is positive (i.e., increases from left to right) or a negative value if the slope of the regression line is negative (i.e., decreases from left to right). The correlation coefficient will always have a value between –1 and +1. The closer the R^2 value is to –1 or +1, the better is the correlation between the two variables being graphically compared. If all the points of a plot fall exactly on a regression line having a positive slope, R^2 will have a value of +1. If all the points of a plot fall exactly on a regression line having a negative slope, R^2 will have a value of –1. If there is no correlation between the two variables (i.e., if the points on a graph appear random), R^2 will be close to 0.

In the following procedure, the Microsoft Excel program will be used to convert the lengths of the100-bp size marker fragments into log values. These values will be graphed on the *y* axis against their migration distance on the *x* axis. Using linear regression, a line of best fit will be drawn through these points and an equation describing that line will be derived. The unknown fragment's migration distance will be placed into the equation for the regression line to determine a log value for the fragment's size. Taking the antilog of this value will yield the unknown fragments size in base pairs.

<table>
<tr>
<td>1. On a computer equipped with Microsoft Office, open the Excel program by clicking on its icon. This will open a spreadsheet for entering data.</td>
<td>
</td>
</tr>
<tr>
<td>2. In column A, row 1, enter "bp." In column B, row 1, enter "Distance (x)." In column C, row 1, enter "log bp (y)." In column 1 beneath the "bp" heading, enter the size of each fragment of the 100-bp ladder, starting with the 1500-bp fragment entered into box A2. In the rows in column B beneath the "Distance (x)" heading, enter the distance, in cm, that each band of the 100-bp ladder migrated from the well.</td>
<td>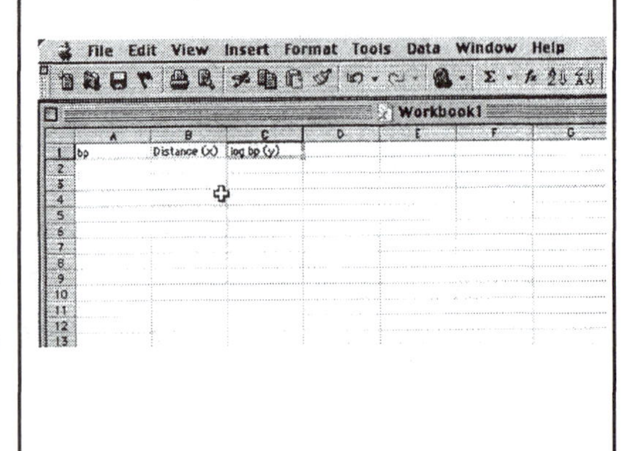
</td>
</tr>
</table>

3. Click on the first box under the "log bp (y)" heading to highlight it. Click the "*fx*" function icon in the toolbar at the top of the spreadsheet to bring up the "Paste Function" box. In the "Function category:" list, select "Math & Trig." In the "Function name:" list, select "LOG 10." Click "OK." The "LOG 10" dialogue box will appear.	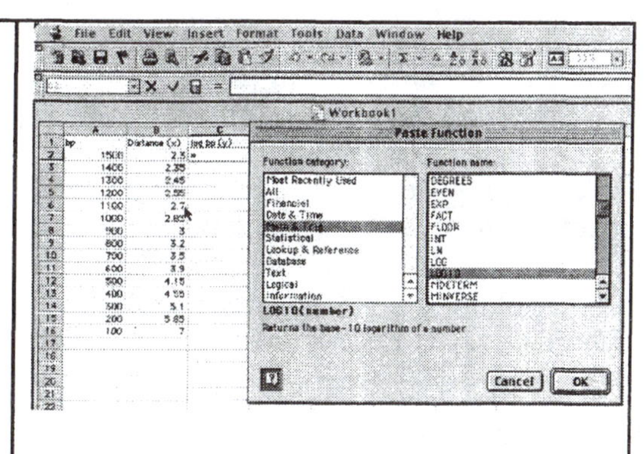
4. Click anywhere within the "LOG 10" dialogue box and drag it out onto the spreadsheet so that the column entries can be seen. Click on 1500 in box A2. "A2" will then appear in the "Number" box. Click "OK." The log value of 1500 will appear in the highlighted box in the "log bp (y)" column. Highlight the entire C column from position "C2" to the last entry in the column. On the keyboard, press "Control" and "D." This will fill down all remaining log values in Column C.	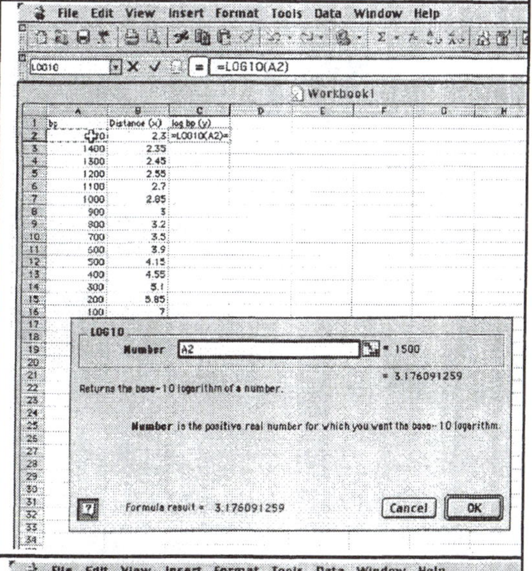
5. Without selecting row 1 (the column heading row) or column 1 (the "bp" column), highlight the values in the "Distance (x)" and "log bp (y)" columns.	

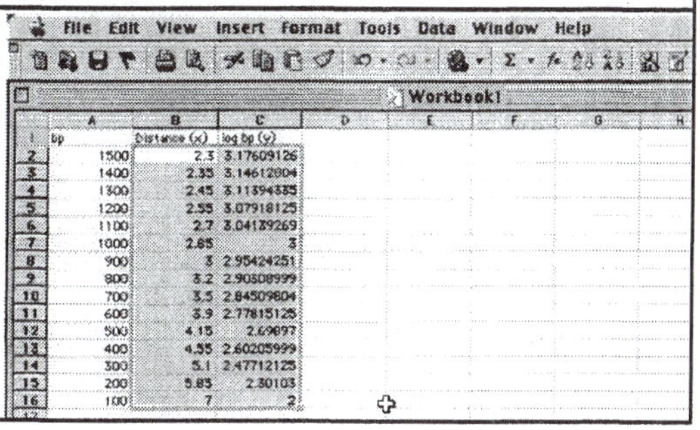

<table>
<tr>
<td>6. In the menu bar, click on the Chart Wizard icon button (it has the appearance of a bar graph) to bring up the Chart Wizard-Chart Type window. Select the "XY (Scatter)" chart type and then click "Finish." A chart will appear over the spreadsheet showing the data plotted on a graph.</td>
<td>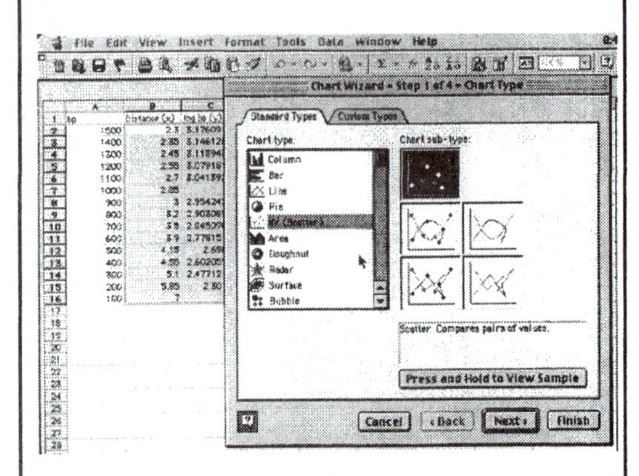
</td>
</tr>
<tr>
<td>7. From the menu bar at the top of the page, pull down the "Chart" list and select "Add Trendline." In the popup window that appears, make sure the "Linear Trend/Regression type" is selected.</td>
<td>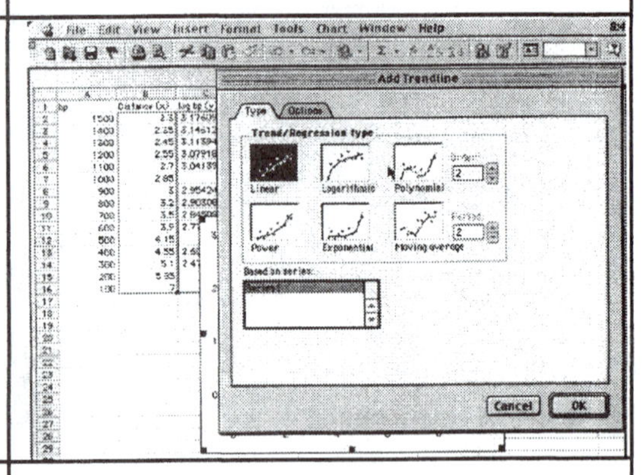
</td>
</tr>
<tr>
<td>8. Click on the "Options" tab in the "Add Trendline" window. In the "Options" window that appears, check the boxes next to "Display equation on chart" and "Display r-squared value on chart." Click "OK." The line of best fit will be drawn on the graph. The regression line equation and the r-squared value will appear at the bottom of the regression line. You can use "Chart Options…" under the "Chart" pull-down menu to label the chart axes and to attach a chart title. (Note: the chart must be selected on the spreadsheet to gain access to the "Chart" pull-down menu.)</td>
<td>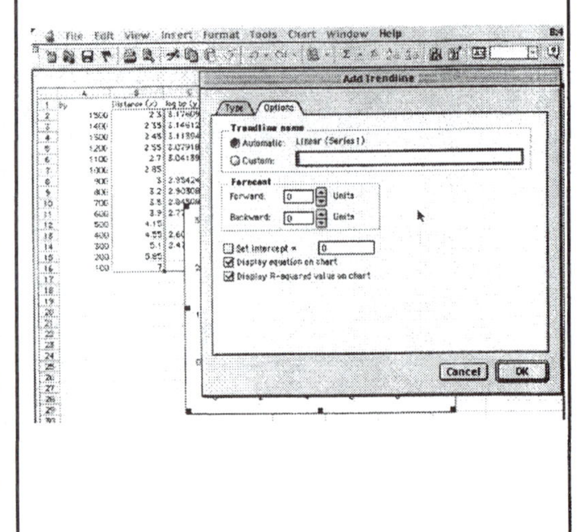
</td>
</tr>
</table>

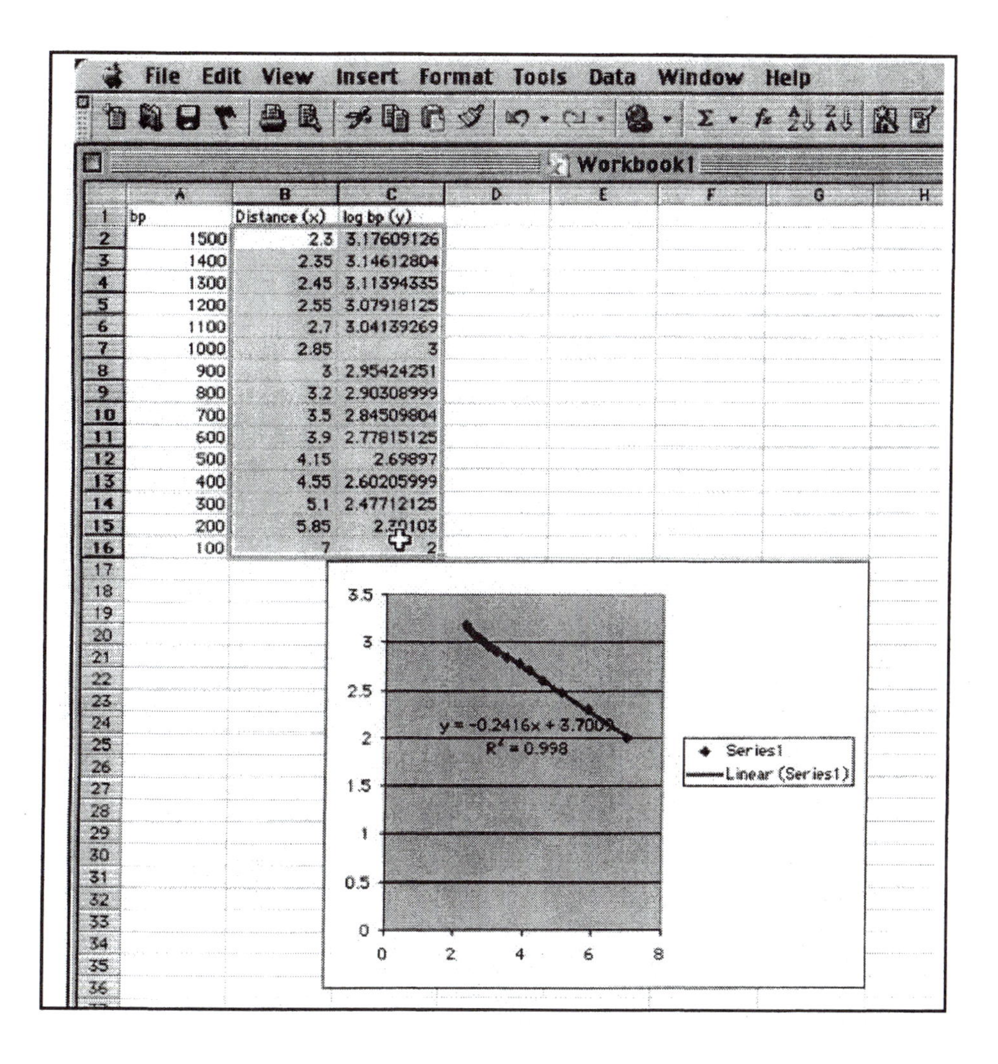

	A	B	C
1	bp	Distance (x)	log bp (y)
2	1500	2.3	3.17609126
3	1400	2.35	3.14612804
4	1300	2.45	3.11394335
5	1200	2.55	3.07918125
6	1100	2.7	3.04139269
7	1000	2.85	3
8	900	3	2.95424251
9	800	3.2	2.90308999
10	700	3.5	2.84509804
11	600	3.9	2.77815125
12	500	4.15	2.69897
13	400	4.55	2.60205999
14	300	5.1	2.47712125
15	200	5.85	2.30103
16	100	7	2

The equation for the regression line is $y = -0.2416x + 3.7009$. The line has an R^2 value of 0.998. All the points on the graph, therefore, lie very close to or exactly on the regression line; there is a strong linear relationship between distance from the well and the log of fragment size.

The unknown fragment was measured to be 4.2 cm from the well. This represents the x variable in the regression equation. The equation for calculating y, the log of fragment size, is then

$$y = -0.2416(4.2) + 3.7009$$

To solve for y using Excel, the following steps should be performed:

1. In a new column on the spreadsheet, enter 4.2, the distance the unknown fragment migrated from the well.
2. Highlight the empty box in the column adjacent to the "4.2" entry.
3. Click the function ("*fx*") button in the toolbar.
4. In the "Paste Function" dialogue box that appears, select "Math & Trig" under "Function category:" and "PRODUCT" under "Function name:." Click the "OK" button.
5. In the "Product Dialogue" box that appears, enter –0.2416 (the regression equation's m value) in the "Number 1" box and 4.2 (the x value) in the "Number 2" box. Click "OK." The product of mx (–1.0147) will be entered in the highlighted box in the spreadsheet.
6. Highlight the box to the right of this new entry.
7. Click the function ("*fx*") button in the toolbar.
8. In the "Paste Function" dialogue box that appears, select "Math & Trig" under "Function category:" and "SUM" under "Function name:." Click the "OK" button.
9. In the dialogue box that appears, enter –1.0147 (the mx product calculated in Step 5) in the "Number 1" box. Enter 3.7009 (the regression equation's b value) in the "Number 2" box. Click "OK." The sum of the entered numbers (2.6862) will appear in the highlighted box.

The y value for the regression equation is 2.6862. It is the logarithm of the fragment size (log bp). To determine the size of the unknown fragment in base pairs, the antilog of this value must be determined. The following steps in Excel will provide the antilog value.

1. Highlight an empty box in a new column to the right of the log value (2.6862) calculated earlier.
2. Click the function ("*fx*") button in the top toolbar.
3. In the "Paste Function" dialogue box that appears, select "Math & Trig" under "Function category:" and "POWER" under "Function name:." Click the "OK" button.
4. In the dialogue box that appears, enter "10" in the "Number" box (since the operation is in base 10). In the "Power" box, enter "2.6862" (the log bp value calculated earlier). Click the "OK" button.
5. The size of the band (in bp) will appear in the highlighted box. The unknown fragment has a size of 485.5 bp.

Note that linear regression gave a size of 485.5 bp, whereas graphing on a semilog plot gave a size of 470 bp. Given the possible error involved in measuring bands on a photograph and possible migration anomalies due to salt or base sequence effects, neither answer is necessarily more accurate.

Generating Nested Deletions Using Nuclease BAL 31

Once a recombinant clone carrying the sought-after gene has been identified, it is often desirable to create deletions of the insert fragment. This can facilitate at least three objectives.

1. Creating a deletion set of the insert fragment can assist in the construction of a restriction map.
2. DNA sequencing of the entire insert fragment can be achieved by preparing a series of deletions that stretch further and further into the insert.
3. The beginning of the inserted gene (if working from a genomic source) can be brought closer to the control elements of the vector.

BAL 31 is an exonuclease that degrades both 3′ and 5′ termini of duplex DNA. The enzyme also has endonuclease activity and cleaves at nicks, gaps, and single-stranded sections of duplex DNA and RNA. Digestion of linear DNA with BAL 31 produces truncated molecules having blunt ends or ends having short overhanging 5′ termini.

The most frequently used method for generating deletion mutants using BAL 31 in a recombinant plasmid is first to cut it at a unique restriction site and then to digest nucleotides away from the ends. The enzyme's activity is dependent on the presence of calcium. The addition of EGTA will chelate the calcium and stop the reaction. The extent of digestion, therefore, can be controlled by adding the chelating agent after defined intervals of time.

The incubation time required to produce the desired deletion can be estimated using the equation of Legersky et al. (1978):

$$M_t = M_0 - \frac{2M_n V_{\max} t}{\left[K_m + (S)_0\right]}$$

where

M_t = molecular weight of the duplex DNA after t minutes of incubation
M_0 = original molecular weight of the duplex DNA
M_n = average molecular weight of a mononucleotide [taken as 330 daltons (Da)]
V_{max} = maximum reaction velocity (expressed as moles of nucleotide removed/liter/minute)
t = length of time of incubation, in minutes
K_m = Michaelis–Menten constant (expressed as moles of double-stranded DNA termini/liter)
S_0 = moles of double-stranded DNA termini/liter at $t = 0$ minutes

Maniatis et al. (1982) recommend using a V_{max} value of 2.4×10^{-5} moles of nucleotide removed /liter/minute when the reaction contains 40 units of BAL 31/mL. They also

make the assumption that the value of V_{max} is proportional to the amount of enzyme in the reaction. For example, if 0.8 units are used in a 100-µ
L reaction, the concentration, in units/mL, becomes

$$\frac{0.8 \text{ units}}{100 \ \mu\text{L}} \times \frac{1000 \ \mu\text{L}}{\text{mL}} = 8 \text{ units/mL}$$

The V_{max} value is proportional to this amount, as follows:

$$\frac{8 \text{ units/mL}}{40 \text{ units/mL}} = \frac{x \text{ mol/L/min}}{2.4 \times 10^{-5} \text{ mol/L/min}}$$

Solving for x yields

$$\frac{(8 \text{ units/mL})(2.4 \times 10^{-5} \text{ mol/L/min})}{40 \text{ units/mL}} = x$$

$$\frac{1.92 \times 10^{-4} \text{ mol/L/min}}{40} = x = 4.8 \times 10^{-6} \text{ mol/L/min}$$

Therefore, if using a BAL 31 concentration of 8 units/mL in the reaction, the new V_{max} value you should use is 4.8×10^{-6} mol/L/min.

Maniatis et al. (1982) also recommend using a K_m value of 4.9×10^{-9} mol of double-stranded DNA termini/liter.

Problem 9.20 Into a 100-µL reaction, you have placed 1 unit of BAL 31 nuclease and 5 µg of a 15,000-bp fragment of DNA. You wish to digest 5000 bp from the termini. How many minutes should you allow the reaction to proceed before stopping it with EGTA?

Solution 9.20 We need to determine values for M_0 (the initial molecular weight of the 15,000-bp fragment), M_t (the molecular weight of the 10,000-bp fragment made by removing 5000 bp from the 15,000-bp original fragment), V_{max} (the new value for this enzyme concentration), and S_0 (the number of double-stranded termini present in 5 µg of a 15,000-bp fragment).

The molecular weight of a 15,000-bp fragment of duplex DNA can be calculated as follows (assuming an average molecular weight of 660 g/mol/bp):

$$15{,}000 \text{ bp} \times \frac{660 \text{ g/mol}}{\text{bp}} = 9.9 \times 10^{6} \text{ g/mol}$$

Therefore, a 15,000-bp fragment has a molecular weight of 9.9×10^6 g/mol. This is equivalent to 9.9×10^6 daltons. This is the M_0 value.

The M_t value, the molecular weight of the 10,000-bp desired product, is then

$$10{,}000 \text{ bp} \times \frac{660 \text{ g/mol}}{\text{bp}} = 6.6 \times 10^6 \text{ g/mol}$$

Therefore, the desired 10,000-bp product has a molecular weight of 6.6×10^6 g/mol, which is equivalent to 6.6×10^6 daltons.

To calculate the new V_{max} value for the enzyme amount used in this problem, we first need to determine the concentration of BAL 31 in the reaction in units/mL. We have 1 unit in a 100-µL reaction. This converts to units/mL as follows:

$$\frac{1 \text{ unit}}{100 \text{ µL}} \times \frac{1000 \text{ µL}}{\text{mL}} = \frac{10 \text{ units}}{\text{mL}}$$

The new V_{max} value can then be calculated by a relationship of ratios:

$$\frac{10 \text{ units/mL}}{40 \text{ units/mL}} = \frac{x \text{ mol/L/min}}{2.4 \times 10^{-5} \text{ mol/L/min}}$$

Solving for x gives

$$x \text{ mol/L/min} = \frac{(10 \text{ units/mL})(2.4 \times 10^{-5} \text{ mol/L/min})}{40 \text{ units/mL}} = 6 \times 10^{-6} \text{ mol/L/min}$$

Therefore, the V_{max} value for this problem is 6×10^{-6} mol/L/min.

The S_0 value (the number of double-stranded termini present in 5 µg of a 15,000-bp fragment) is calculated in several steps. We have determined that a 15,000-bp fragment has a molecular weight of 9.9×10^6 g/mol. We can now determine how many moles of this 15,000-bp fragment are in 5 µg (the amount of DNA added to our reaction):

$$5 \text{ µg} \times \frac{1 \text{ g}}{1 \times 10^6 \text{ µg}} \times \frac{1 \text{ mol}}{9.9 \times 10^6 \text{ g}} = \frac{5 \text{ mol}}{9.9 \times 10^{12}} = 5.1 \times 10^{-13} \text{ mol}$$

Since we are using a 15,000-bp linear fragment, there are two termini per fragment, or 1×10^{-12} mol of double-stranded DNA termini:

$$2 \times (5.1 \times 10^{-13} \text{ mol}) = 1 \times 10^{-12} \text{ mol termini}$$

To obtain the S_0 value, we need to determine how many moles of termini there are per liter. Since our reaction has a volume of 100 μL, this converts to moles per liter as follows:

$$\frac{1\times10^{-12}\ \text{mol termini}}{100\ \mu\text{L}}\times\frac{1\times10^{6}\ \mu\text{L}}{\text{L}}=\frac{1\times10^{-8}\ \text{mol termini}}{\text{L}}$$

Therefore, the S_0 value is 1×10^{-8} moles termini/L. We now have all the values needed to set up the equation. The original expression is

$$M_t = M_0 - \frac{2M_nV_{\max}t}{\left[K_m+(S)_0\right]}$$

Substituting the given and calculated values yields the following equation:

$$6.6\times10^{6}\ \text{Da}=9.9\times10^{6}\ \text{Da}-\frac{2(330\ \text{Da})\left(6\times10^{-6}\ \text{mol / L / min}\right)t\ \text{min}}{\left(\frac{4.9\times10^{-9}\ \text{mol termini}}{\text{L}}\right)+\left(\frac{1\times10^{-8}\ \text{mol termini}}{\text{L}}\right)}$$

Simplifying the equation gives the following result:

$$6.6\times10^{6}\ \text{Da}=9.9\times10^{6}\ \text{Da}-\frac{\left(3.96\times10^{-3}\ \text{Da}\right)t}{1.49\times10^{-8}}$$

$$6.6\times10^{6}\ \text{Da}=9.9\times10^{6}\ \text{Da}-(2.66\times10^{5}\ \text{Da})t$$

Subtracting 9.9×10^{6} from each side of the equation yields

$$-3.3\times10^{6}\ \text{Da}=(-2.66\times10^{5}\ \text{Da})t$$

Dividing each side of the equation by -2.66×10^{5} yields the following result:

$$\frac{-3.3\times10^{6}\ \text{Da}}{-2.66\times10^{5}\ \text{Da}}=t$$

$$t=12.4\ \text{minutes}$$

Therefore, to remove 5000 bp from a 15,000-bp linear fragment in a 100-μL reaction containing 1 unit of BAL 31 and 5 μg of 15,000-bp fragment, allow the reaction to proceed for 12.4 minutes before adding EGTA.

References

Clark, L., and J. Carbon (1976). A colony bank containing synthetic ColE1 hybrid plasmids representative of the entire *E. coli* genome. *Cell* 9:91.

Davis, L.G., M.D. Dibner, and J.F. Battey (1986). *Basic Methods in Molecular Biology*. Elsevier Science, New York.

Dugaiczyk, A., H.W. Boyer, and H.M. Goodman (1975). Ligation of *Eco*R I endonuclease-generated DNA fragments into linear and circular structures. *J. Mol. Biol.* 96:174–184.

Laird, C.D. (1971). Chromatid structure: relationship between DNA content and nucleotide sequence diversity. *Chromosoma* 32:378.

Legersky, R.J., J.L. Hodnett, and H.B. Gray, Jr. (1978). Extracellular nucleases of pseudomonad *Bal*31 III. Use of the double-strand deoxyriboexonuclease activity as the basis of a convenient method for mapping fragments of DNA produced by cleavage with restriction enzymes. *Nucl. Acids Res.* 5:1445.

Maniatis, T., E.F. Fritsch, and J. Sambrook (1982). *Molecular Cloning: A Laboratory Manual*. Cold Spring Harbor Laboratory, Cold Spring Harbor, N.Y.

Wallace, B.R. and C.G. Miyada (1987). Oligonucleotide probes for the screening of recombinant DNA libraries. *Meth. Enzymol.* 152:432.

Wallace B.R., J. Shaffer, R.C. Murphy, J. Bonner, T. Hirose, and K. Itakura (1979). Hybridization of synthetic oligodeoxyribonucleotides to φX174 DNA: the effect of single base pair mismatch. *Nucl. Acids Res.* 6:3543.

Protein

10

Introduction

There are a number of reasons for creating recombinant DNA clones. One is to test the activity of a foreign promoter or enhancer control element. This is accomplished by placing the control element in close proximity to a reporter gene encoding an easily assayed enzyme. The most popular genes for this purpose are the *lacZ* gene encoding β-galactosidase, the *cat* gene encoding chloramphenicol acetyltransferase (CAT), and the *luc* gene encoding the light-emitting protein luciferase.

The amount of enzyme produced in a reporter gene system is usually expressed in terms of units. For most proteins, a **unit** is defined as the amount of enzyme activity per period of time divided by the quantity of protein needed to give that activity under defined reaction conditions in a specified volume. A unit of the restriction endonuclease *Eco* RI, for example, is defined as the amount of enzyme required to completely digest 1 μg of double-stranded substrate DNA in 60 minutes at 37°C in a 50-μL reaction volume. To calculate units of protein, therefore, requires the assay of both protein activity and protein quantity.

This chapter addresses the calculation needed to quantitate protein and, for the cases of several examples, to assess protein activity.

Protein Quantitation by Measuring Absorbance at 280 nm

There are a number of methods for estimating protein concentration. One of the simplest is to measure absorbance at 280 nm. Proteins absorb ultraviolet light at 280 nm primarily because of the presence of the ringed amino acids tyrosine and tryptophan. Since different proteins have different amounts of tyrosine and tryptophan residues, there will be variablity in the degree to which different proteins absorb light at 280 nm. In addition, conditions that alter a protein's tertiary structure (buffer type, pH, and reducing agents) can affect its absorbance. Despite the variability, reading absorbance at 280 nm is often used because few other chemicals also absorb at this wavelength.

As a very rough approximation, 1 absorbance unit at 280 nm is equal to 1 mg protein/mL. The protein solution being assayed should be diluted if its absorbance at 280 nm is greater than 2.0.

Problem 10.1 A solution of a purified protein is diluted 0.2 mL into a total volume of 1.0 mL in a cuvette having a 1-cm light path. A spectrophotometer reading at 280 nm gives a value of 0.75. What is a rough estimate of the protein concentration?

Solution 10.1 This problem can be solved by setting up a relationship of ratios as follows:

$$\frac{1 \text{ mg/mL}}{1 \text{ absorbance unit}} = \frac{x \text{ mg/mL}}{0.75 \text{ absorbance units}}$$

Solving for x yields the following result:

$$\frac{(1 \text{ mg/mL}) \times (0.75)}{1} = x = 0.75 \text{ mg protein/mL}$$

Since the protein sample was diluted 0.2 mL/1.0 mL, to determine the protein concentration of the original sample, this result must be multiplied by the inverse of the dilution factor:

$$\frac{0.75 \text{ mg}}{\text{mL}} \times \frac{1.0 \text{ mL}}{0.2 \text{ mL}} = 3.75 \text{ mg protein/mL}$$

Therefore, the purified protein solution has an approximate concentration of 3.75 mg/mL.

Using Absorbance Coefficients and Extinction Coefficients to Estimate Protein Concentration

The **absorbance coefficient** of a protein is its absorbance for a certain concentration at a given wavelength in a 1-cm cuvette. Absorbance coefficients are usually determined for convenient concentrations, such as 1 mg/mL, 1%, or 1 molar. In the *CRC Handbook of Biochemistry and Molecular Biology* (CRC Press, Boca Raton, FL), absorbance coefficients are given for 1% solutions ($A^{1\%}$). The absorbance coefficient of a 1% protein solution is equivalent to the absorbance of a

1% solution of a given protein at a given wavelength in a 1-cm cuvette. The absorbance coefficient of a 1 mg/mL protein solution (written $A^{1\text{mg/mL}}$) is equivalent to the absorbance of a 1 mg/mL solution of a given protein at a given wavelength in a 1-cm cuvette. Since absorbance coefficients are dependent on pH and ionic strength, it is important when using the published values to duplicate the protein's environment under which the coefficients were determined.

The **molar extinction coefficient** (E_M) of a protein is equivalent to the absorbance of a 1 *M* solution of a given protein at a given wavelength in a 1-cm cuvette.

Using the coefficients, protein concentration can be determined using the following relationships (Stoscheck, 1990):

$$\text{Protein concentration (in mg/mL)} = \frac{\text{Absorbance}}{A_{1\text{cm}}^{1\text{mg/mL}}}$$

$$\text{Protein concentration (in \%)} = \frac{\text{Absorbance}}{A_{1\text{cm}}^{1\%}}$$

$$\text{Protein concentration (in molarity)} = \frac{\text{Absorbance}}{E_M}$$

Problem 10.2 Acetylcholinesterase from *E. electricus* has an absorbance coefficient ($A^{1\%}$) of 22.9 in 0.1 *M* NaCl, 0.03 *M* sodium phosphate, pH 7.0 at 280 nm. A solution of this protein gives a reading at 280 nm of 0.34 in a 1-cm cuvette. What is its percent concentration?

Solution 10.2 The following equation will be used:

$$\text{Protein concentration} = \frac{\text{Absorbance}}{A_{1\text{cm}}^{1\%}}$$

Substituting the values provided gives the following result:

$$\text{\% protein concentration} = \frac{0.34}{22.9} = 0.015\%$$

Therefore, a solution of acetylcholinesterase having an absorbance of 0.34 has a concentration of 0.015%.

Problem 10.3 Acetylcholinesterase has a molar extinction coefficient (E_M) of 5.27×10^5. In a 1-cm cuvette, a solution of the protein has an A_{280} of 0.22. What is its molar concentration?

Solution 10.3 The following equation will be used:

$$\text{Protein concentration (in molarity)} = \frac{\text{Absorbance}}{E_M}$$

Substituting the values provided gives the following result:

$$M = \frac{0.22}{5.27 \times 10^5} = 4.2 \times 10^{-7} \ M$$

Therefore, the solution of acetylcholinesterase has a concentration of 4.2×10^{-7} *M*.

Relating Absorbance Coefficient to Molar Extinction Coefficient

The absorbance coefficient for a1% protein solution ($A_{1cm}^{1\%}$) is related to the molar extinction coefficient (E_M) in the following way (Kirschenbaum, 1976):

$$E_M = \frac{(A_{1cm}^{1\%})(\text{molecular weight})}{10}$$

Problem 10.4 β-Galactosidase from *E. coli* has a molecular weight of 750,000. Its absorbance coefficient for a 1% solution is 19.1. What is its molar extinction coefficient?

Solution 10.4 Substituting the values provided into the preceding equation gives the following result:

$$E_M = \frac{(19.1)(750,000)}{10} = \frac{1.4 \times 10^7}{10} = 1.4 \times 10^6$$

Therefore, β-galactosidase has a molar extinction coefficient of 1.4×10^6.

Determining a Protein's Extinction Coefficient

A protien's extinction coefficient at 205 nm can be determined using the following formula (Scopes, 1974):

$$E_{205\text{nm}}^{1\text{mg/mL}} = 27 + 120\left(\frac{A_{280}}{A_{205}}\right)$$

This equation can be used when the protein has an unknown concentration. Once a concentration has been determined, the extinction coefficient at 280 nm can be determined using the following equation.

$$E_{280\text{ nm}} = \frac{\text{mg protein/mL}}{\text{A}_{280}}$$

Note*: This equation will not give an accurate result for proteins having an unusual phenylalanine content.*

Problem 10.5 A solution of protein having an unknown concentration is diluted 30 μL into a final volume of 1 mL. At 205 nm, the diluted solution gives an absorbance of 0.648. Absorbance at 280 nm is 0.012. What is the molar extinction coefficient at 205 nm and at 280 nm?

Solution 10.5 The given values are substituted into the equation for determining a protein's extinction coefficient.

$$E_{205\text{nm}}^{1\text{mg/mL}} = 27 + 120\left(\frac{A_{280}}{A_{205}}\right)$$

$$E_{205\text{nm}} = 27 + 120\left(\frac{0.012}{0.648}\right)$$

$$= 27 + 120(0.019)$$

$$= 27 + 2.28 = 29.28$$

Therefore, the extinction coefficient at 205 nm is 29.28 for the diluted sample. To obtain the extinction coefficient for the undiluted sample, this value is multiplied by the dilution factor:

$$29.28 \times \frac{1000\ \mu\text{L}}{30\ \mu\text{L}} = 976$$

Therefore, the molar extinction coefficient at 205 nm is 976.

The protein concentration can now be determined using the earlier equation relating molarity, absorbance, and the molar extinction coefficient:

$$\text{Molarity} = \frac{A_{205}}{E_{205}}$$

Since the molar extinction coefficient, $E_{205\text{nm}}$, was calculated for the sample (not the diluted sample), the A_{205} value must be multiplied by the dilution factor so that all terms are determined from equivalent data:

$$A_{205} = 0.648 \times \frac{1000\ \mu\text{L}}{30\ \mu\text{L}} = 21.6$$

The equation for calculating molarity is then written as follows:

$$\text{Molarity} = \frac{21.6}{976} = 0.22\ M$$

The molar extinction coefficient at 280 nm ($E_{280\text{nm}}$) is now calculated using the following relationship:

$$E_{280\text{nm}} = \frac{\text{Molarity}}{A_{280}}$$

To use this equation for this problem, the A_{280} value must first be multiplied by the dilution factor so that all terms are treated equivalently:

$$A_{280} = 0.012 \times \frac{1000\ \mu\text{L}}{30\ \mu\text{L}} = 0.4$$

The molar extinction coefficient at 280 nm can now be calculated.

$$E_{280\text{nm}} = \frac{0.022\ M}{0.4} = 0.055$$

Therefore, the molar extinction coefficient for this protein at 205 nm is 976 and at 280 nm is 0.4.

Relating Concentration in Milligrams per Milliliter to Molarity

Protein concentration in milligrams per milliliter is related to molarity in the following way:

$$\text{Concentration (in mg/mL)} = \text{Molarity} \times \text{Protein molecular weight}$$

This relationship can be used to convert a concentration in milligrams per milliliter to molarity, and vice versa. Note that this equation will actually give a result in grams per liter. However, this is the same as milligrams per milliliter. For example, 4 g/L is the same as 4 milligrams per milliliter as shown by the following relationship:

$$\frac{4\ \text{g}}{\text{L}} \times \frac{1\ \text{L}}{1000\ \text{mL}} \times \frac{1000\ \text{mg}}{\text{g}} = \frac{4\ \text{mg}}{\text{mL}}$$

Problem 10.6 A protein has a molecular weight of 135,000 g/mol. What is the concentration of a $2 \times 10^{-4}\ M$ solution in milligrams per milliliter?

Solution 10.6 Using the relationship between concentration in milligrams per milliliter and molarity we have

$$\text{mg/mL} = \frac{2 \times 10^{-4}\ \text{mol}}{\text{L}} \times \frac{135{,}000\ \text{g}}{\text{mol}} \times \frac{1000\ \text{mg}}{\text{g}} \times \frac{1\ \text{L}}{1000\,\text{mL}} = 27\ \text{mg/mL}$$

Therefore, a $2 \times 10^{-4}\ M$ solution of a protein having a molecular weight of 135,000 has a concentration of 27 mg/mL.

Problem 10.7 A protein has a molecular weight of 167,000. A solution of this protein has a concentration of 2 mg/mL. What is the molarity of the solution?

Solution 10.7 To solve for molarity, the equation for converting molarity and molecular weight to milligrams per milliliter must be rearranged as follows:

$$\text{Molarity} = \frac{\text{Concentration (in mg/mL)}}{\text{Protein molecular weight}}$$

Substituting the given values into this equation gives the following result:

$$\text{Molarity} = \frac{2 \text{ mg/mL}}{167{,}000 \text{ g/mol}} = 1.2 \times 10^{-5} \ M$$

Therefore, the protein solution has a concentration of 1.2×10^{-5} *M*.

Protein Quantitation Using A_{280} When Contaminating Nucleic Acids Are Present

Although nucleic acids maximally absorb UV light at 260 nm, they do show strong absorbance at 280 nm. Measuring protein concentration of crude extracts also containing nucleic acids (RNA and DNA) will give artificially high protein estimates. You must therefore correct the reading at 280 by subtracting the contribution from nucleic acids. This relationship is given by the following equation (Layne, 1957):

$$\text{Protein concentration (in mg/mL)} = (1.55 \times A_{280}) - (0.76 \times A_{260})$$

Problem 10.8 An experimenter is to determine the concentration of protein in a crude extract. She zeros the spectrophotometer with buffer at 260 nm, dilutes 10 μL of the extract into a total volume of 1.0 mL in a 1-cm quartz cuvette, and then takes a reading of the crude extract. She obtains an A_{260} reading of 0.075. She zeros the spectrophotometer with buffer at 280 nm and reads the diluted extract at that wavelength. She obtains an A_{280} value of 0.25. What is the approximate concentration of protein in the crude extract?

Solution 10.8 Using the preceding equation and substituting in the values she obtained on the spectrophotometer gives the following result:

$$\text{mg/mL} = (1.55 \times 0.25) - (0.76 \times 0.075)$$

$$= 0.39 - 0.06 = 0.33 \text{ mg/mL}$$

Therefore, the protein in the 1-mL cuvette has an approximate concentration of 0.33 mg/mL. The experimenter, however, diluted her sample 10 μL/1000 μL to take a measurement. To determine the protein concentration of the undiluted sample, this value must be multiplied by the inverse of the dilution factor:

$$0.33 \text{ mg/mL} \times \frac{1000 \ \mu\text{L}}{10 \ \mu\text{L}} = 33 \text{ mg/mL}$$

Therefore, the original, undiluted sample has an approximate protein concentration of 33 mg/mL.

Protein Quantitation at 205 nm

Proteins also absorb light at 205 nm and with greater sensitivity than at 280 nm. Absorbance at this wavelength is due primarily to the peptide bonds between amino acids. Since all proteins have peptide bonds, there is less variability between proteins at the 205-nm wavelength than at the 280-nm wavelength. Whereas quantitation using 280 nm works best for protein in the range of 0.02 mg to 3 mg in a 1-mL cuvette, absorbance at 205 nm is more sensitive and works for 1 to 100 μg of protein per milliliter.

Protein concentration at 205 nm can be estimated using the equation from Stoscheck (1990):

$$\text{Protein concentration (in mg/mL)} = 31 \times A_{205}$$

Problem 10.9 A sample of purified protein is diluted 2 μL into a final volume of 1.0 mL in a quartz cuvette. A reading at 205 nm gives a value of 0.0023. What is the approximate protein concentration of the original sample?

Solution 10.9 Using the preceding equation with these values gives

$$\text{mg/mL} = 31 \times 0.146 = 4.53 \text{ mg/mL}$$

Therefore, the concentration of protein in the cuvette is 0.07 mg/mL. However, this represents the concentration of the diluted sample. To obtain the protein concentration of the original, undiluted sample, we must multiply this value by the inverse of the dilution factor:

$$4.53 \text{ mg/mL} \times \frac{1000 \ \mu\text{L}}{2 \ \mu\text{L}} = 2265 \text{ mg/mL}$$

Therefore, the undiluted sample has a concentration of 2265 mg protein/mL.

Protein Quantitation at 205 nm When Contaminating Nucleic Acids Are Present

Nucleic acids contribute very little to an absorbance reading at 205 nm. However, they do contribute more substantially at the other wavelength where proteins absorb, 280 nm. To correct for the contribution from RNA and DNA to protein estimation in crude extracts, the following equation is used (Peterson, 1983):

$$\text{Protein concentration (in mg/mL)} = \frac{A_{205}}{\left(27 + (120)\frac{A_{280}}{A_{205}}\right)}$$

Problem 10.10 A crude extract is diluted 2.5 μL into a final volume of 1000 μL in a 1-cm cuvette. Absorbancy readings at 205 nm and 280 nm are taken, with the following results:

$$A_{205} = 0.648$$

$$A_{280} = 0.086$$

What is the concentration of protein in the crude extract?

Solution 10.10 Using the preceding formula and substituting the values provided gives the following result:

$$\text{mg/mL} = \frac{0.648}{\left[27 + (120)\frac{0.086}{0.648}\right]} = \frac{0.648}{27 + (120)0.13} = \frac{0.648}{27 + 15.6} = \frac{0.648}{42.6} = 0.015 \text{ mg/mL}$$

Therefore, the concentration of protein in the cuvette is 0.015 mg/mL. However, since this is a diluted sample, to obtain the protein concentration in the original, undiluted sample, this value must be multiplied by the inverse of the dilution factor:

$$0.015 \text{ mg/mL} \times \frac{1000 \ \mu\text{L}}{2.5 \ \mu\text{L}} = 6 \text{ mg/mL}$$

Therefore, the crude extract has a protein concentration of approximately 6 mg/mL.

Measuring Protein Concentration by Colorimetric Assay – The Bradford Assay

There are several published methods for measuring protein concentration by colorimetric assay. Each of these relies on an interaction between the protein and a chemical reagent dye. This interaction leads to a color change detectable by a spectrophotometer. The absorbance of an unknown protein sample is compared to a standard curve generated by a protein diluted in a series of known concentrations. The concentration of the unknown sample is interpolated from the standard curve.

One of the most popular methods for quantitating proteins by a dye reagent is called the Bradford Assay [after its developer, Marion Bradford (Bradford, 1976)]. The method uses Coomassie Brilliant Blue G-250 as a protein-binding dye. When protein binds to the dye, an increase in absorbance at 595 occurs. The dye binding reaction takes only a couple of minutes and the color change is stable for up to an hour. The assay is sensitive down to approximately 200 μg protein/mL.

Problem 10.11 An experimenter prepares dilutions of bovine serum albumin (BSA) in amounts ranging from 10 μg to 100 μg in a volume of 0.1 mL (adjusted to volume with buffer where necessary). Five mL of dye reagent (Bradford, 1976) is added to each tube, the mixture is vortexed and allowed to sit for at least 2 minutes at room temperature, and the absorbance at 595 nm is read on a spectrophotometer. The results shown in Table 10.1 are obtained.

µg BSA	A_{595}
10	0.11
20	0.21
30	0.30
40	0.39
50	0.48
60	0.58
75	0.71
100	0.88

Table 10.1 Absorbance at 595 nm for samples of BSA having known protein amounts.

A purified protein of unknown concentration is diluted 2 µL into a total volume of 100 µL and reacted with the Bradford dye reagent in the same manner as the protein standard. Its absorbance at 595 nm is 0.44. What is the concentration of protein in the original, undiluted sample, in milligrams per milliliter?

Solution 10.11 The data is plotted as a standard curve (Figure 10.1). To determine the amount of protein in the sample of unknown concentration, a horizontal line from 0.44 absorbance units on the *y* axis is drawn to the standard curve. A vertical line from the point of interception is then drawn down to the *x* axis (see upcoming Figure 10.2). This vertical line intercepts the *x* axis at 46 µg. Therefore, in the assay reaction tube, there are 46 µg of protein, or, thought of another way, the 2 µL taken from the protein prep into the assay tube contained 46 µg of protein. This represents a concentration of 23 mg/mL in the original sample, as shown in the following conversion:

$$\frac{46\ \mu\text{g}}{2\ \mu\text{L}} \times \frac{1000\ \mu\text{L}}{1\ \text{mL}} \times \frac{1\ \text{mg}}{1000\ \mu\text{g}} = \frac{23\ \text{mg}}{\text{mL}}$$

Therefore, the original, undiluted protein preparation has a concentration of 23 mg protein/mL.

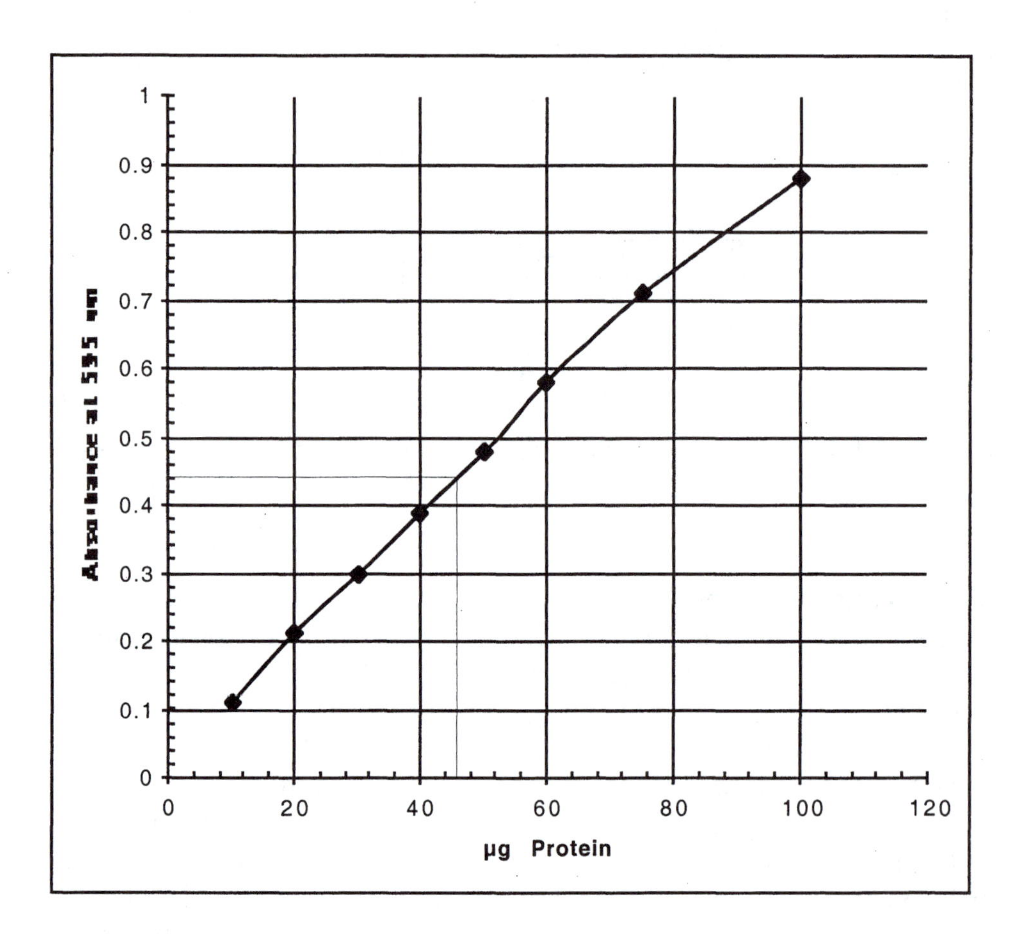

Figure 10.1 Data plot for Problem 10.11.

Using β-Galactosidase to Monitor Promoter Activity and Gene Expression

The sugar lactose is a disaccharide of galactose and glucose joined by a β-galactoside bond. The enzyme β-galactosidase, the product of the *lacZ* gene, breaks the β-galactoside bond to release the galactose and glucose monosaccharides. A compound known as ONPG (*o*-nitrophenol-β-D-galactopyranoside) has a structure that mimics lactose (Figure 10.2) and can be cleaved by β-galactosidase. ONPG, however, carries an *o*-nitrophenol ring group rather than glucose. When ONPG, a colorless compound, is cleaved, *o*-nitrophenol (ONP), a yellow compound, is released. ONP absorbs light at 420 nm. This provides a convenient assay for β-galactosidase activity. In the presence of excess ONPG, the amount of yellow color (i.e., the amount of *o*-nitrophenol released) is directly proportional to the amount of β-galactosidase in the assay and to the time allowed for the reaction. Because β-galactosidase is so easily assayed, the *lacZ* gene can be placed downstream from a promoter to determine its activity under different conditions. The *lacZ* gene can also be fused in reading frame to the gene of another protein under the control of its natural promoter to monitor expression of that promoter.

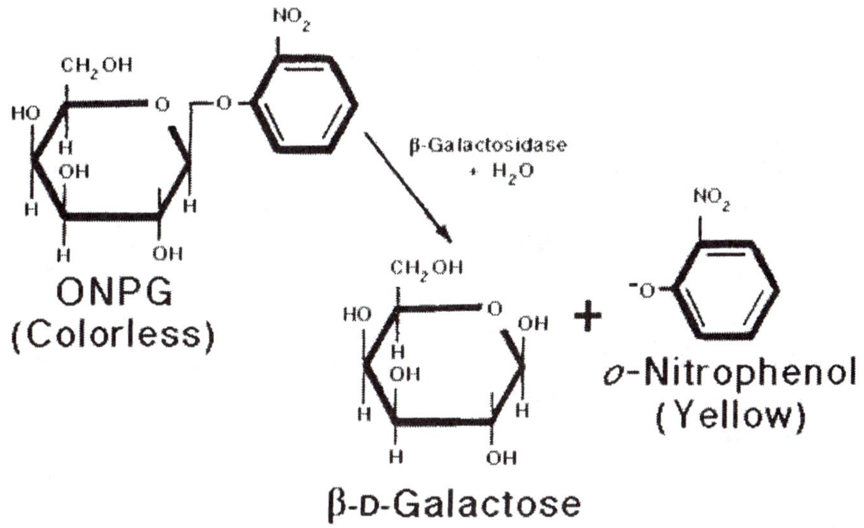

Figure 10.2 β-Galactosidase splits ONPG, a colorless compound, into galactose and *o*-nitrophenol. The *o*-nitrophenol released is yellow and absorbs light at 420 nm.

Assaying β-Galactosidase in Cell Culture

β-Galactosidase is typically assayed from *E. coli* cells growing in culture. Cells are grown to log phase ($2–5 \times 10^8$ cells/mL; OD_{600} of 0.28 to 0.70), the culture is cooled on ice for 20 minutes to prevent further growth, and 1 mL is withdrawn for an OD_{600} reading. An aliquot of the culture is added to a phosphate buffer (Z buffer) containing KCl, magnesium sulfate, and β-mercaptoethanol such that the final volume of the assay reaction is 1 mL. To make the cells permeable to ONPG, a drop of toluene is added to the reaction tube and it is then vortexed for 10 seconds. The toluene is allowed to evaporate and the assay tube is then placed in a 28°C waterbath for 5 minutes. The reaction is initiated by adding ONPG. The reaction time is monitored, and when sufficient yellow color has developed, the reaction is stopped by the addition of sodium carbonate to raise the pH to a level at which β-galactosidase activity is inactivated. The reaction should take between 15 minutes and 6 hours to produce sufficient yellow color. The optical density is then measured at 420 nm and 550 nm. The optimal 420 nm reading should be between 0.6 and 0.9. If it does not fall within this range, the sample should be diluted appropriately.

Both the *o*-nitrophenol and light scatter from cell debris contribute to the absorbance at 420 nm. Light scatter is corrected for by taking an absorbance measurement at 550 nm, a wavelength at which there is no contribution from *o*-nitrophenol. The amount of light scattering at 420 nm is proportional to the amount at 520 nm. This relationship is described by the equation

$$\text{OD}_{420}\text{ light scattering} = 1.75 \times \text{OD}_{550}$$

The absorbance of only the *o*-nitrophenol component at 420 nm can be computed by using this relationship as a correction factor and subtracting the OD_{550} multiplied by 1.75 from the 420-nm absorbance reading. Units of β-galactosidase are then calculated using the formula of Miller (1972):

$$\text{Units} = 1000 \times \frac{\text{OD}_{420} - 1.75 \times \text{OD}_{550}}{t \times v \times \text{OD}_{600}}$$

where OD_{420} and OD_{550} are read from the reaction mixture, OD_{600} is the absorbance of the cell culture just before assay, t is equal to the reaction time (in minutes), and v is the volume (in mL) of aliquoted culture used in the assay.

Problem 10.12 The *lacZ* gene (β-galactosidase) has been cloned adjacent to an inducible promoter. Recombinant cells are grown to log phase, and a sample is withdrawn to measure β-galactosidase levels during the uninduced state. Inducer is then added and the cells are allowed to grow for another 10 minutes before being chilled on ice. A 1-mL aliquot of the culture has an OD_{600} of 0.4. A 0.2-mL aliquot of the culture is added to 0.8 mL of Z buffer, toluene, and ONPG. After 42 minutes, Na_2CO_3 is added to stop the reaction. A measurement at 420 nm gives a reading of 0.8. Measurement at 550 nm gives a reading of 0.02. How many units of β-galactosidase are present in the 0.2 mL of induced culture?

Solution 10.12 Using the equation

$$\text{Units} = 1000 \times \frac{OD_{420} - 1.75 \times OD_{550}}{t \times v \times OD_{600}}$$

and substituting our values gives the following result:

$$\text{Units} = 1000 \times \frac{(0.8) - (1.75 \times 0.02)}{42 \times 0.2 \times 0.4} = \frac{765}{3.36}$$

$$= 228 \text{ units enzyme}$$

Therefore, there are 228 units of β-galactosidase in 0.2 mL of culture, or 228 units/0.2 mL = 1140 units β-galactosidase/mL.

Specific Activity

It may sometimes be desirable to calculate the specific activity of β-galactosidase in a reaction. **Specific activity** is defined as the number of units of enzyme per milligram of protein. Protein quantitation was covered earlier in this chapter. However, even without directly quantitating the amount of protein in an extract, an amount can be estimated by assuming that there are 150 μg of protein per 10^9 cells and that, for *E. coli*, an OD_{600} of 1.4 is equivalent to 10^9 cells/mL.

Problem 10.13 If it assumed that 1 mL of cells described in Problem 10.12 contains 43 μg of protein, what is the specific activity of the β-galactosidase for that problem?

Solution 10.13. It was calculated in Problem 10.12 that 1 mL of culture contained 1140 units of β-galactosidase per milliliter. The specific activity is then calculated as follows:

$$\frac{1140 \text{ units}}{43 \text{ μg protein}} \times \frac{1000 \text{ μg}}{\text{mg}} = 26{,}512 \text{ units/mg protein}$$

Therefore, the β-galactosidase in the cell culture has a specific activity of 26,512 units/mg protein.

Assaying β-Galactosidase from Purified Cell Extracts

For some applications, β-galactosidase might be assayed from a more purified source than that described in Problem 10.12. For example, if the cells are removed by centrifugation prior to assay or if the protein (or fusion protein) is being purified by biochemical techniques, then the OD_{600} value that is part of the calculation for determining β-galactosidase units is irrelevant.

In a protocol described by Robertson et al. (1997), β-galactosidase activity (Z_a) is calculated as the number of micromoles of ONPG converted to ONP per unit time and does not require a cell density measurement at 600 nm. It is given by the equation

$$Z_a = \frac{\left(\dfrac{A_{410}}{E_{410}}\right)V}{t}$$

where

A_{410} = absorbance at 410 nm
E_{410} = molar extinction coefficient of ONP at 410 nm (3.5 mL/μmol) assuming the use of a cuvette having a 1-cm light path
V = reaction volume
t = time, in minutes, allowed for the development of yellow color

To perform the β-galactosidase assay, 2.7 mL of 0.1 *M* sodium phosphate, pH 7.3, 0.1 mL of 3.36 *M* 2-mercaptoethanol, 0.1 mL of 30 m*M* $MgCl_2$, and 0.1 mL of 34 m*M* ONPG are combined in a spectrophotometer cuvette and mixed. The reaction mix is zeroed in the spectrophotometer at 410 nm. Ten μL of the sample to be assayed for β-galactosidase activity are added to the 3 mL of reaction mix, and the increase in absorbance at 410 nm is monitored. Monitoring is discontinued when A_{410} has reached between 0.2 and 1.2.

Problem 10.14

(a) A gene having an in-frame *lacZ* fusion is placed under control of the SV40 early promoter on a recombinant plasmid that is transfected into mammalian cells in tissue culture. After 48 hours, 1 mL of cells is sonicated and centrifuged to remove cellular debris. After blanking the spectrophotometer against reagent,

10 μL of sample are added to 3 mL of reagent mix prepared as described earlier, and the increase in absorbance at 410 nm is monitored. After 12 minutes, an absorbance of 0.4 is achieved. What is the β-galactosidase enzyme activity, in micromoles per minute, for the reaction sample?

(b) A 20-μL sample of cleared cell lysate is used in the Bradford protein assay to measure protein quantity. It is discovered that this aliquot contains 180 μg of protein. What is the specific activity of the sample?

Solution 10.14(a) The assay gave the following results:

$$A_{410} = 0.4$$
$$t = 12 \text{ min}$$
$$V = 3.01 \text{ mL (3 mL of reaction mix + 10 μL sample)}$$

These values are placed into the equation for calculating β-galactosidase activity:

$$Z_a = \frac{\left(\frac{A_{410}}{E_{410}}\right)V}{t}$$

$$Z_a = \frac{\left(\frac{0.4}{3.5 \text{ mL/μmol}}\right)3.01 \text{ mL}}{12 \text{ min}}$$

$$Z_a = \frac{(0.114 \text{ μmol})(3.01)}{12 \text{ min}} = \frac{0.343 \text{ μmol}}{12 \text{ min}} = 0.03 \text{ μmol/min}$$

Therefore, the 3.01-mL reaction produced 0.03 μmol of ONP per minute.

Solution 10.14(b) The Bradford assay for protein content gave 180 μg of protein in the 20-μL sample. This is converted to micrograms per milliliter as follows:

$$\frac{180\ \mu\text{g protein}}{20\ \mu\text{L}} = 9\ \mu\text{g}/\mu\text{L}$$

The amount of protein in the 10-μL sample used in the β-galactosidase assay is calculated as follows:

$$10\ \mu\text{L} \times \frac{9\ \mu\text{g protein}}{\mu\text{L}} = 90\ \mu\text{g protein}$$

The specific activity of the 10-μL sample in the enzyme assay is the activity divided by the amount of protein in that sample:

$$\text{Specific activity} = \frac{0.03\ \mu\text{mol/min}}{90\,\mu\text{g}} = \frac{0.0003\ \mu\text{mol/min}}{\mu\text{g}}$$

Therefore, the sample has a specific activity of 3×10^{-4} μmol/min/μg.

The CAT Assay

The *cat* gene, encoding chloramphenicol acetyltransferase (CAT), confers resistance to the antibiotic chloramphenicol to those bacteria that carry and express it. The *cat* gene, having evolved in bacteria, is not found naturally in mammalian cells. However, when introduced into mammalian cells by transfection and when under the control of an appropriate promoter, the *cat* gene can be expressed. It has found popularity, therefore, as a reporter gene for monitoring transfection and for studying gene expression within transfected cells.

Chloramphenicol acetyltransferase transfers an acetyl group ($-COCH_3$) from acetyl coenzyme A to two possible positions on the chloramphenicol molecule. To assay for CAT activity, an extract from cells expressing CAT is incubated with [^{14}C]-labeled chloramphenicol. Transfer of the acetyl group from acetyl coenzyme A to chloramphenicol alters the mobility of the antibiotic on a thin-layer chromatography (TLC) plate. The amount of acetylated chloramphenicol is directly proportional to the amount of CAT enzyme.

Protocols for running a CAT assay have been described by Gorman et al. (1982) and by Kingston (1989). Typically, 1×10^7 mammalian cells are transfected with a plasmid carrying *cat* under a eukaryotic promoter. After two to three days, the cells are resuspended in 100 μL of 250 m*M* Tris, pH 7.5. The cells are sonicated and centrifuged at 4°C to remove debris. The extract is incubated at 60°C for 10 minutes to inactivate endogenous acetylases and spun again at 4°C for 5 minutes. Twenty μL of extract are added to a reaction containing 1 μCi [^{14}C] of chloramphenicol (50 μCi/mmol), 140 m*M* Tris-HCl, pH 7.5, and 0.44 m*M* acetyl coenzyme A in a final volume of 180 μL. The reaction is allowed to proceed at 37°C for 1 hour. Four hundred μL of cold ethyl acetate are added and mixed in, and the reaction is spun for 1 minute. The upper organic phase is transferred to a new tube and dried under vacuum. The dried sample is resuspended in 30 μL of ethyl acetate, and a small volume (5–10 μL) is spotted close to the edge of a silica gel thin-layer chromatography plate. The TLC plate is placed in a chromatography tank pre-equilibrated with 200 mL of chloroform:methanol (19:1). Chromatography is allowed to proceed until the solvent front is two-thirds of the way up the plate. The plate is air-dried, a pen with radioactive ink is used to mark the plate for orientation, and the plate is exposed to X-ray film. Since chroloramphenicol has two potential acetylation sites, two acetylated forms, running faster than the unacetylated form, are produced by the CAT reaction (Figure 10.3). A third, higher spot, corresponding to double acetylation, can be seen under conditions of high CAT activity when single-acetylation sites have been saturated. When this third spot appears, the assay will be out of its linear range and the extract should be diluted or less time allowed for the assay. The area on the TLC plate corresponding to the chloramphenicol spot and the acetylated chloramphenicol spots are scraped off and counted in a liquid scintillation counter. CAT activity is then calculated as follows:

$$\text{CAT activity} = \%\ \text{chloramphenicol aceylated}$$

$$= \frac{\text{Counts in acetylated species}}{\text{Counts in acetylated species} + \text{Nonacetylated chloramphenicol counts}} \times 100$$

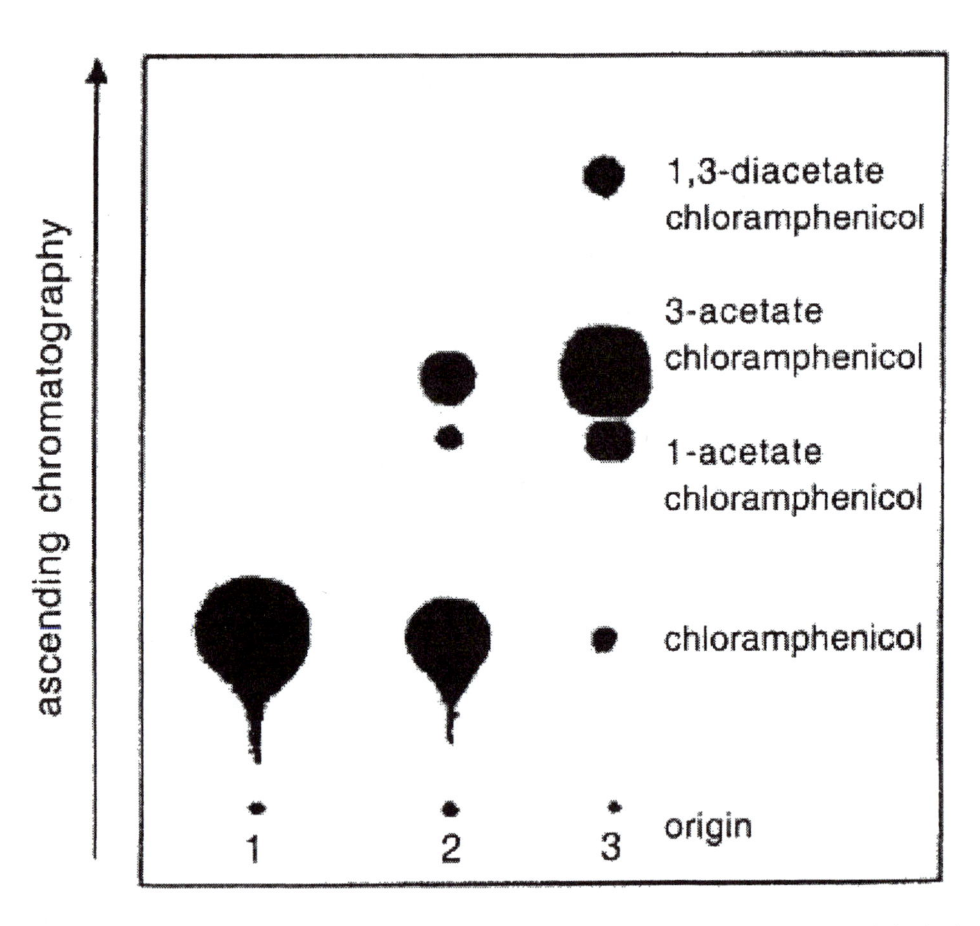

Figure 10.3 TLC plate assay of CAT activity. Spot 1 contains only labeled chloramphenicol. Spot 2 shows the two spots created by single-chloramphenicol-acetylation events. Spot 3 shows the spot toward the top of the plate representing double-acetylation events.

Problem 10.15 A transfection experiment using CAT as a reporter is conducted as just described. The acetylated species of chloramphenicol gave 3×10^5 cpm. The unacetylated chloramphenicol spot gave 1.46×10^6 cpm. What is the CAT activity?

Solution 10.15 The values obtained in the liquid scintillation counter are placed into the formula for calculating CAT activity, as follows:

CAT activity = % chloramphenicol aceylated

$$= \frac{\text{Counts in acetylated species}}{\text{Counts in acetylated species + Nonacetylated chloramphenicol counts}} \times 100$$

$$\%\ \text{CAT activity} = \frac{3 \times 10^5\ \text{cpm}}{3 \times 10^5\ \text{cpm} + 1.46 \times 10^6\ \text{cpm}} \times 100 = 17\%$$

Therefore, 17% of the chloramphenicol was acetylated by CAT.

Calculating Molecules of CAT

Purified CAT protein has a specific activity of 153,000 nmol/min/mg of protein (Shaw and Brodsky, 1968) and has a molecular weight of 25,668 g/mol (Shaw et al., 1979). These values can be used to calculate the number of molecules of CAT in an assay, as shown in the next problem.

Problem 10.16 For Problem 10.15, how many molecules of CAT are contained in the extract?

Solution 10.16 The CAT assay contaned 1 μCi of [^{14}C] chloramphenicol having a specific activity of 50 μCi/mmol. The amount of chloramphenicol in the reaction is then calculated as follows:

$$1\ \mu\text{Ci} \times \frac{\text{mmol}}{50\ \mu\text{Ci}} = 0.02\ \text{mmol}$$

Seventeen percent of the [^{14}C] chloramphenicol was converted to the acetylated forms during the assay. The number of millimoles of chloramphenicol acetylated by CAT is then calculated by multiplying 0.02 mmol by 17/100 (17%):

$$0.02\ \text{mmol} \times \frac{17}{100} = 0.0034\ \text{mmol}$$

The reaction in the assay was incubated at 37°C for 60 minutes. The amount of [^{14}C] chloramphenicol acetylated per minute is calculated as follows:

$$\frac{0.0034 \text{ mmol}}{60 \text{ min}} = 5.7 \times 10^{-5} \text{ mmol/min}$$

Since the specific activity of CAT is given in units of nanomoles, a conversion factor must be used to convert millimoles to nanomoles. Likewise, since it also contains a "milligram" term and since molecular weight is defined as grams/mole, another conversion factor will be used to convert milligrams to grams. Avogadro's number (6.023×10^{23} molecules per mole) will be used to calculate the final number of molecules. The expression for calculating the number of molecules of CAT in the assay can be written as follows:

$$\frac{5.7 \times 10^{-5} \text{ mmol}}{\text{min}} \times \frac{1 \times 10^{6} \text{ nmol}}{\text{mmol}} \times \frac{\text{mg}}{153,000 \text{ nmol/min}} \times \frac{\text{g}}{1 \times 10^{3} \text{ mg}} \times \frac{1 \text{ mol}}{25,668 \text{ g}} \times \frac{6.023 \times 10^{23} \text{ molecules}}{\text{mol}}$$

Note that the term

$$\frac{\text{mg}}{153,000 \text{ nmol/min}}$$

in the equation is the same as

$$\text{mg} \times \frac{\text{min}}{153,000 \text{ nmol}}$$

All units cancel, therefore, except "molecules," the one term desired. Multiplying numerator and denominator values gives the following result:

$$= \frac{3.4 \times 10^{25} \text{ molecules}}{3.9 \times 10^{12}} = 8.7 \times 10^{12} \text{ molecules}$$

Therefore, the reaction in Problem 10.15 contains 8.7×10^{12} molecules of CAT.

Use of Luciferase in a Reporter Assay

The *luc* gene encoding the enzyme luciferase from the firefly *Photinus pyralis* has become a popular reporter for assaying gene expression. Luciferase from transformed bacterial cells or transfected mammalian cells expressing the *luc* gene is assayed by placing cell extract in a reaction mix containing excess amounts of luciferin (the enzyme's substrate), ATP, $MgSO_4$, and molecular oxygen (O_2). ATP-dependent oxidation of luciferin leads to a burst of light that can be detected and quantitated (as light units) at 562 nm in a luminometer. The amount of light generated in the reaction is proportional to luciferase concentration. The luciferase assay is more sensitive than that for either β-galactosidase or CAT.

Because the luminometer used for assaying luciferase activity provides a reading in light units (LU's), very little math is required to evaluate the assay. In the following problem, however, the number of molecules of luciferase enzyme in a reporter gene assay will be determined.

Problem 10.17 To derive a standard curve, different amounts of purified firefly luciferase are assayed for activity in a luminometer. A consistent peak emission of 10 light units is the lower level of activity reliably detected by the instrument, and this corresponds to 0.35 pg of enzyme. How many molecules of luciferase does this represent?

Solution 10.17 Luciferase has a molecular weight of 60,746 g/mol (De Wet et al., 1987). The number of molecules of luciferase is calculated by multiplying the protein quantity by its molecular weight and Avogadro's number. A conversion factor is used to convert picograms to grams. The calculation is as follows:

$$0.35 \text{ pg} \times \frac{1 \text{ g}}{1 \times 10^{12} \text{ pg}} \times \frac{1 \text{ mol}}{60{,}746 \text{ g}} \times \frac{6.023 \times 10^{23} \text{ molecules}}{\text{mol}} = \frac{2.11 \times 10^{23} \text{ molecules}}{6.075 \times 10^{16}}$$

$$= 3.47 \times 10^{6} \text{ molecules}$$

Therefore, 0.35 pg of luciferase corresponds to 3.47×10^6 molecules.

In Vitro Translation – Determining Amino Acid Incorporation

In vitro translation can be used to radiolabel a protein from a specific gene. A small amount of the tagged protein can then be added as a tracer to follow the purification of the same protein from a cell extract. As for any experiment using radioactive labeling, it is important to determine the efficiency of its incorporation into the target molecule.

To prepare an in vitro translation reaction, the gene to be expressed should be under the control of an active promoter (*lac*UV5, *tac*, lambda P_L, etc.). The DNA carrying the desired gene is placed in a reaction containing S30 or S100 extract freed of exogenous DNA and mRNA (Spirin *et al.*, 1988). The reaction should also contain triphosphates, tRNAs, ATP, GTP, and a mixture of the amino acids in which one of them (typically methionine) is added as a radioactive compound.

To determine the amount of incorporation of the radioactive amino acid, an aliquot of the reaction is precipitated with cold TCA and casamino acids (as carrier). The precipitate is collected on a glass fiber filter, rinsed with TCA several times, and then rinsed a final time with acetone to remove unincorporated [^{35}S] methionine. The filter is dried and counted in a liquid scintillation counter.

Problem 10.18 A gene is transcribed and translated in a 100-µL reaction. Into that reaction are added 2 µL of 20 mCi/mL [^{35}S] methionine (2000 Ci/mmol). After incubation at 37°C for 60 minutes, a 4-µL aliquot is TCA precipitated to determine incorporation of radioactive label. In the liquid scintillation counter, the filter gives off 1.53×10^6 cpm. A 2-µL aliquot of the reaction is counted directly in the scintillation counter to determine total counts in the reaction. This sample gives 2.4×10^7 cpm. What percent of the [^{35}S] methionine has been incorporated into polypeptide?

Solution 10.18 To solve this problem, the total number of counts in the reaction must be determined. The number of counts in the aliquot from the reaction precipitated by TCA will then be calculated and the number of counts expected in the entire translation reaction will be determined. The percent [^{35}S] methionine incorporated will then be calculated as a final step.

Two μL of the reaction were placed directly into scintillation fluid and counted. Since 2 μL gave 2.4×10^7 cpm and there are a total of 100 μL in the reaction, the total number of counts in the reaction is calculated as follows:

$$\frac{2.4\times10^7\ \text{cpm}}{2\ \mu\text{L}}\times100\ \mu\text{L}=1.2\times10^9\ \text{cpm}$$

Therefore, a total of 1.2×10^9 cpm are in the 100-μL reaction.

Four μL were treated with TCA. On the filter, 1.53×10^6 cpm were precipitated. The total number of counts expected in the 100-μL reaction is then determined as follows:

$$\frac{1.53\times10^6\ \text{cpm}}{4\ \mu\text{L}}\times100\ \mu\text{L}=3.83\times10^7\ \text{cpm}$$

Therefore, a total of 3.83×10^7 cpm would be expected to be incorporated in the reaction.

The percent total radioactivity incorporated into protein is then calculated as the total number of counts incorporated divided by the total number of counts in the reaction, multiplied by 100:

$$\frac{3.83\times10^7\ \text{cpm}}{1.2\times10^9\ \text{cpm}}\times100=3.2\%$$

Therefore, by this TCA precipitation assay, 3.2% of the [^{35}S] methionine was incorporated into polypeptide.

References

Bradford, M. M. (1976). A rapid and sensitive method for the quantitation of microgram quantities of protein utilizing the principle of protein-dye binding. *Analyt. Biochem.* 72:248.

De Wet, J.R., K.V. Wood, M. DeLuca, D.R. Helinski, and S. Subramani (1987). Firefly luciferase gene: structure and expression in mammalian cells. *Mol. Cell Biol.* 7:725–737.

Gorman, C.M., L.F. Moffat, and B.H. Howard (1982). Recombinant genomes which express chloramphenicol acetyltransferase in mammalian cells. *Molec. Cell. Biol.* 2:1044–1051.

Kingston, R.E. (1989) Uses of fusion genes in mammalian transfection: harvest and assay for chloramphenicol acetyltransferase. In *Short Protocols in Molecular Biology: A Compendium of Methods from Current Protocols in Molecular Biology* (Ausubel, F.M., R. Brent, R.E. Kingston, D.D. Moore, J.G. Seidman, J.A. Smith, K. Struhl, P. Wang-Iverson, S.G. Bonitz, eds.), Wiley, New York, pp. 268–270.

Kirschenbaum, D.M. (1976). In *CRC Handbook of Biochemistry and Molecular Biology* (G.D. Fasman, ed.), 3rd ed., Vol. 2, p. 383. CRC Press, Boca Raton, FL.

Layne, E. (1957). Spectrophotometric and turbidimetric methods for measuring proteins. *Meth. Enzymol.* 3:447–454.

Miller, J.H. (1972). *Experiments in Molecular Genetics.* Cold Spring Harbor Laboratory, Cold Spring Harbor, NY, pp. 352–355.

Peterson, G.L. (1983). Determination of total protein. *Meth. Enzymol.* 91:95–119.

Robertson, D., S. Shore, and D.M. Miller (1997). *Manipulation and Expression of Recombinant DNA: A Laboratory Manual.* Academic Press, San Diego, CA, pp. 133–138.

Scopes, R.K. (1974). Measurement of protein by spectrophotometry at 205 nm. *Analyt. Biochem.* 59:277–282.

Shaw, W.V., and R.F. Brodsky (1968). Characterization of chloramphenicol acetyltransferase from chloramphenicol-resistant *Staphylococcus aureus. J. Bacteriol.* 95:28–36.

Shaw, W.V., L.C. Packman, B.D. Burleigh, A. Dell, H.R. Morris, and B.S. Hartley (1979). Primary structure of a chloramphenicol acetyltransferase specified by R plasmids. *Nature* 282:870-872.

Spirin, A.S., V.I. Baranov, L.A. Ryabova, S.Y. Ovodov, and Y.B. Alakhov (1988). A continuous cell-free translation system capable of producing polypeptides in high yield. *Science* 242:1162.

Stoscheck, C.M. (1990). Quantitation of protein. *Meth. Enzymol.* 182:50–68.

Centrifugation 11

Introduction

Centrifugation is a method of separating molecules having different densities by spinning them in solution around an axis (in a centrifuge rotor) at high speed. It is one of the most useful and frequently employed techniques in the molecular biology laboratory. Centrifugation is used to collect cells, to precipitate DNA, to purify virus particles, and to distinguish subtle differences in the conformation of molecules. Most laboratories undertaking active research will have more than one type of centrifuge, each capable of using a variety of rotors. Small tabletop centrifuges can be used to pellet cells or to collect strands of DNA during ethanol precipitation. Ultracentrifuges can be used to band plasmid DNA in a cesium chloride gradient or to differentiate various structures of replicating DNA in a sucrose gradient.

Relative Centrifugal Force (*g* Force)

During centrifugation, protein or DNA molecules in suspension are forced to the furthest point from the center of rotation. The rate at which this occurs depends on the speed of the rotor (measured as **rpm**, revolutions per minute). The force generated by the spinning rotor is described as the **relative centrifugal force** (**RCF**, also called the "***g***" force) and is proportional to the square of the rotor speed and the radial distance (the distance of the molecule being separated from the axis of rotation). The RCF is calculated using the equation

$$\text{RCF} = 11.17(r)\left(\frac{\text{rpm}}{1000}\right)^2$$

where r is equal to the radius from the centerline of the rotor to the point in the centrifuge tube where the RCF value is needed (in centimeters) and rpm is equal to the rotor speed in revolutions per minute.

Because the RCF increases with the square of the rotor speed, doubling the rpm will increase the RCF four fold.

The RCF is usually expressed in units of gravitational force written as "$\times g$" (times the force of gravity). One g is equivalent to the force of gravity. Two g's is twice the force of gravity. An RCF of 12,500 is 12,500 times the force of gravity and is written as 12,500 $\times g$.

As the distance from the center of the rotor increases, so does the centrifugal force. Most rotor specifications from the manufacturer will give the rotor's radius (***r***) in three measurements. The **minimum radius** (***r*$_{\mathbf{min}}$**) is the distance from the rotor's centerline to the top of the liquid in the centrifuge tube when placed in the rotor at its particular angle (see Figure 11.1). The **maximum radius** (***r*$_{\mathbf{max}}$**) is the distance from the rotor centerline to the furthest point in the centrifuge tube away from the axis of rotation (Figure 11.1). The **average radius** (***r*$_{\mathbf{avg}}$**) is equal to the minimum radius plus the maximum radius divided by 2.

$$r_{ave} = \frac{r_{min} + r_{max}}{2}$$

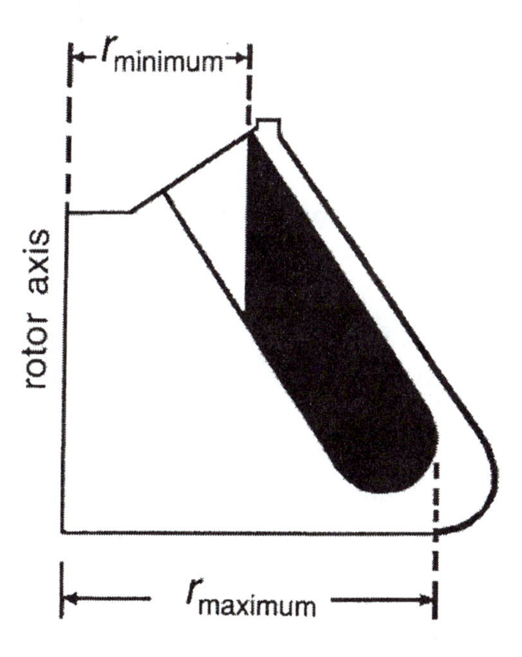

Figure 11.1 Cross section of a typical centrifuge rotor showing the minimum and maximum radii as measured from the rotor axis.

Since the minimum radius and the average radius will vary depending on the volume of liquid being centrifuged, RCF is usually calculated using the maximum radius value.

Problem 11.1 *E. coli* cells are to be pelleted in an SS-34 rotor (maximum radius of 10.7 cm) by centrifugation at 7000 rpm. What is the RCF (*g* force)?

Solution 11.1 We will use the following equation:

$$\text{RCF} = 11.17(r)\left(\frac{\text{rpm}}{1000}\right)^2$$

Substituting the values provided gives the following result:

$$\text{RCF} = 11.17(10.7)\left(\frac{7000}{1000}\right)^2$$

$$\text{RCF} = (119.52)(7)^2 = (119.52)(49) = 5856$$

Therefore, 7000 rpm in an SS-34 rotor is equivalent to 5856 $\times g$.

Converting *g* Force to Revolutions per Minute

The majority of protocols written in the "Materials and Methods" sections of journal articles will give a description of centrifugation in terms of the *g* force. This allows other researchers to reproduce the laboratory method even if they have a different type of centrifuge and rotor. To solve for rpm when the *g* force (RCF) is given, the equation

$$\text{RCF} = 11.17(r)\left(\frac{\text{rpm}}{1000}\right)^2$$

becomes

$$\text{rpm} = 1000\sqrt{\frac{\text{RCF}}{11.17(r)}}$$

Problem 11.2 A protocol calls for centrifugation at 6000 $\times g$. What rpm should be used with an SS-34 rotor (maximum radius of 10.7 cm) to attain this g force?

Solution 11.2 Placing our values into the preceding equation gives

$$\text{rpm} = 1000\sqrt{\frac{6000}{11.17(10.7)}}$$

$$\text{rpm} = 1000\sqrt{\frac{6000}{119.52}} = 1000\sqrt{50.2} = 1000(7.1) = 7100$$

Therefore, to spin a sample at 6000 $\times g$ in an SS-34 rotor, set the centrifuge for 7100 rpm.

Determining *g* Force and Revolutions per Minute by Use of a Nomogram

We have determined g force and rpm values using the following equation:

$$g = 11.17(r)\left(\frac{\text{rpm}}{1000}\right)^2$$

If a large number of g values are calculated using a large number of different values for rpm and spinning radii, a chart approximating the following figure would be generated (Figure 11.2). This chart, called a **nomogram**, is a graphical representation of the relationship between rpm, g force, and a rotor's radius. If you know any two values, the third can be determined by use of a straightedge. The straightedge is lined up such that it intersects the two known values on their respective scales. The third value is given by the intersection of the line on the third scale.

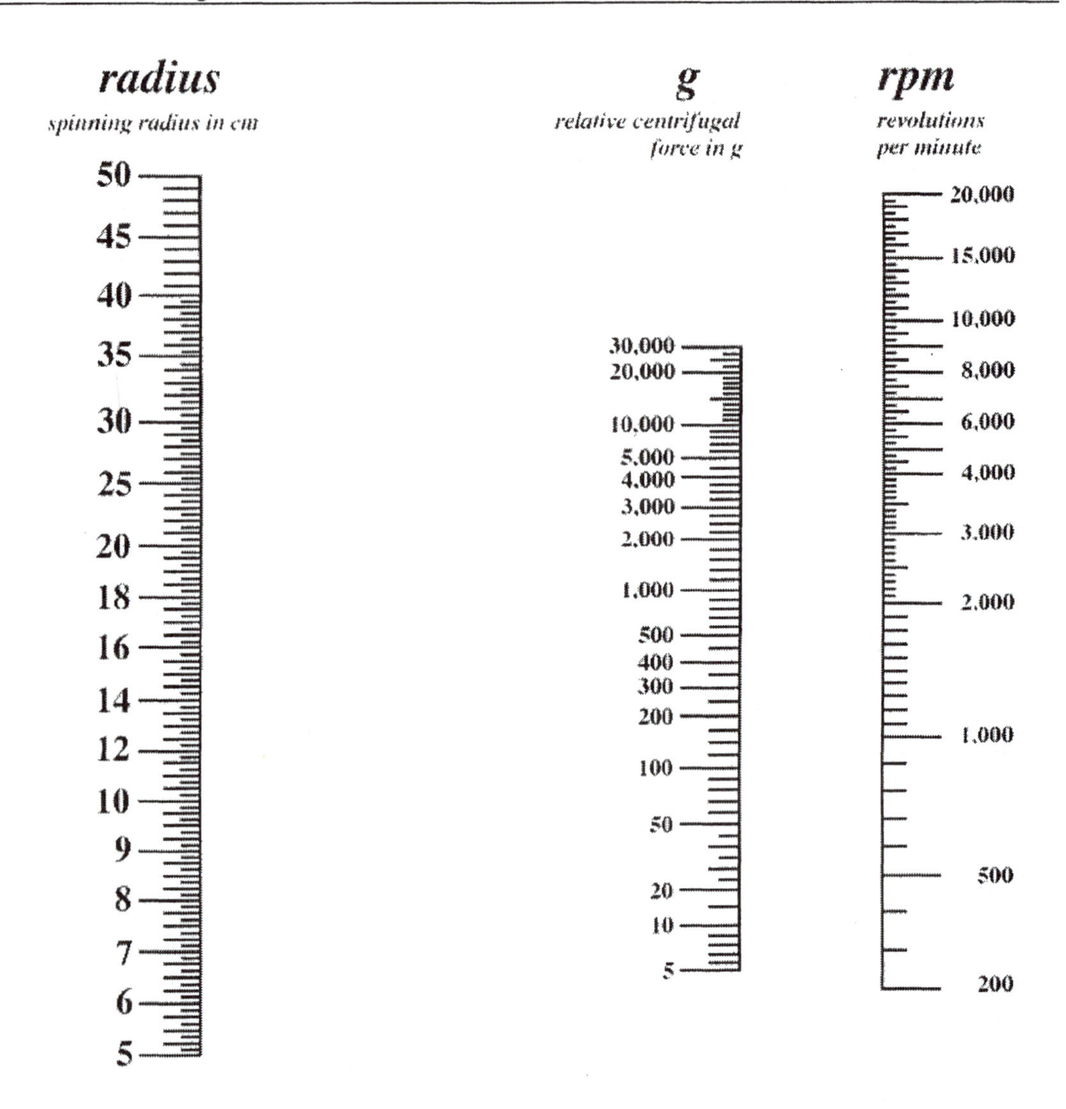

Figure 11.2 Nomogram for determining centrifugation parameters.

Problem 11.3 A rotor with a 7.5-cm spinning radius is used to pellet a sample of cells. The cells are to be spun at 700 $\times g$. What rpm should be used?

Solution 11.3 Using the nomogram in Figure 11.2, a straight line is drawn connecting the three scales such that it passes through both the 7.5-cm mark on the "radius" scale and the 700 mark on the "*g*" scale. This line will intersect the "rpm" scale at 3000 (Figure 11.3). Therefore, the rotor should be spun at

3000 rpm.

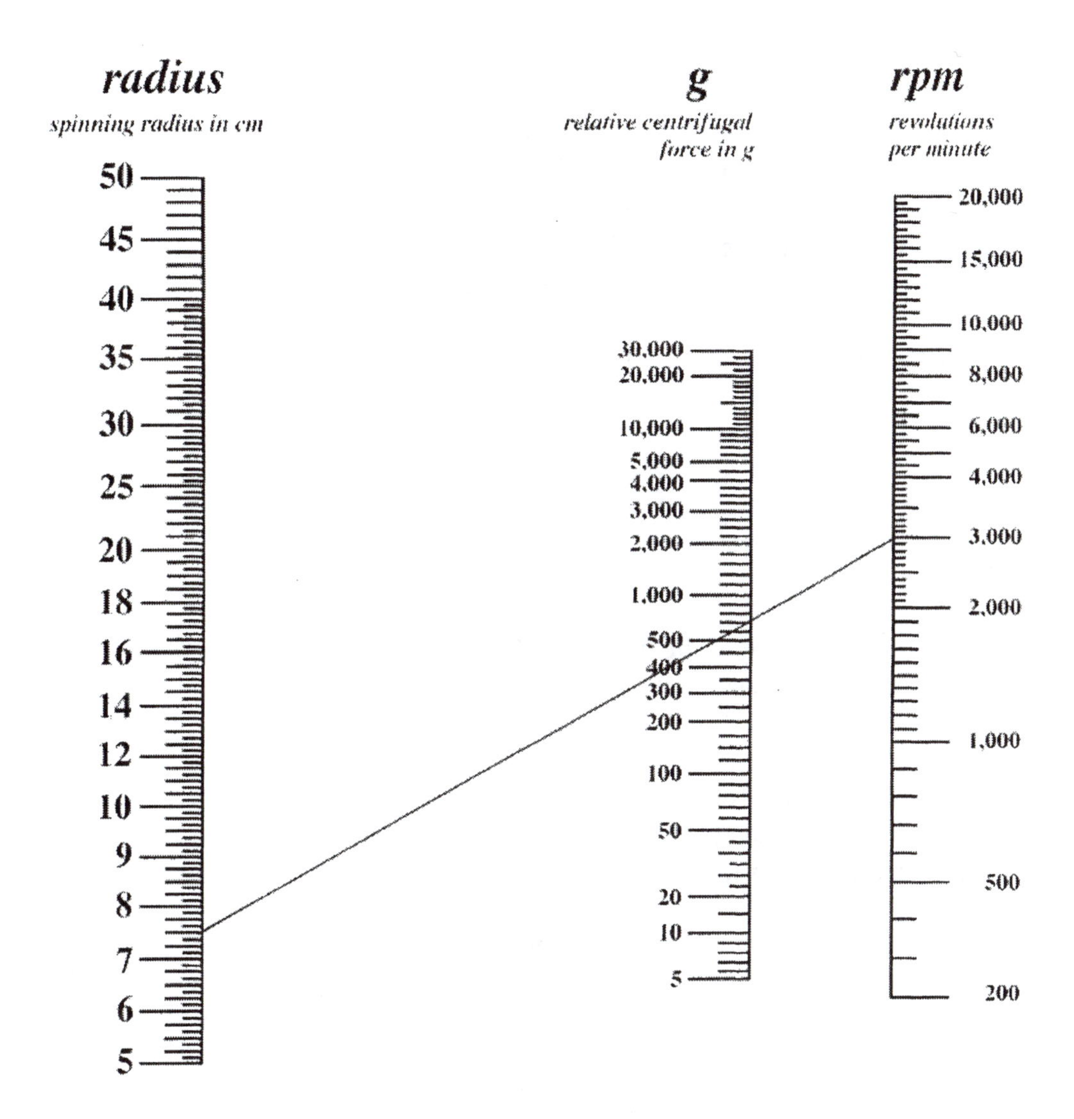

Figure 11.3 Solution to Problem 11.3.

Calculating Sedimentation Times

The time (in hours) required to pellet a particle in water at 20°C through the maximum rotor path length (the distance between r_{min} and r_{max}) is given by the equation

$$t = \frac{k}{s}$$

where

t = sedimentation time in hours
k = clearing factor for the rotor
s = sedimentation coefficient of the particle being pelleted in water at 20°C as expressed in Svedberg units (see later).

The clearing factor, k, is defined by the equation

$$k = \frac{(253{,}000)\left[\ln\left(\frac{r_{max}}{r_{min}}\right)\right]}{\left(\frac{\text{rpm}}{1000}\right)^2}$$

The k factor is a measure of the pelleting efficiency of the rotor. Its value decreases with increasing rotor speed. The lower the k factor, the more efficient is the rotor's ability to pellet the desired molecule. The k factor for any rotor, as provided by the manufacturer, assumes that the fluid being centrifuged has the density of water at 20°C. The k factor will increase if the medium has a higher density or viscosity than water. For example, at the same rotor speed, the k factor is higher for a sucrose gradient than for a water medium.

Several factors will affect the rate at which a particular biological molecule will pellet during centrifugation. Its size, shape, and buoyancy all influence the rate at which a molecule passes through the medium. A molecule having a high molecular weight will sediment faster than one having a low molecular weight. A **Svedberg unit** describes the sedimentation rate of a molecule and is related to its size. It is expressed as a dimension of time such that 1 Svedberg unit is equal to 10^{-13} seconds. For most biological molecules, the Svedberg unit value lies between 1S (1 Svedberg unit, 1×10^{-13} seconds) and 200S (200 Svedberg units, 200×10^{-13} seconds). The names of the ribosomal subunits 30S and 50S from *E. coli,* for example, derive from their sedimentation rates as expressed in Svedberg units. The same is true for the bacterial ribosomal RNAs 5S, 16S, and 23S.

Problem 11.4 An SS-34 rotor has an r_{min} of 3.27 cm and an r_{max} of 10.70 cm for a sealed tube containing the maximum allowable volume. If a 120S molecule is to be spun in an aqueous medium at 20,000 rpm at 20°C, how long will it take to pellet the molecule?

Solution 11.4 The *k* factor must first be calculated. Substituting the given values into the given equation yields the following result:

$$k = \frac{(253{,}000)\left[\ln\left(\frac{10.70}{3.27}\right)\right]}{\left(\frac{20{,}000}{1000}\right)^2}$$

$$k = \frac{(253{,}000)[\ln(3.27)]}{(20)^2} = \frac{(253{,}000)(1.18)}{400} = \frac{298{,}540}{400} = 746$$

Therefore, an SS-34 rotor spinning at 20,000 rpm has a *k* factor value of 746. Placing this value into the equation for calculating sedimentation time gives

$$t = \frac{749}{120\text{S}} = 6.24 \text{ hours} = 6 \text{ hours, } 14 \text{ minutes}$$

Therefore, to pellet the 120S molecule, it must be spun in an SS-34 rotor at 20,000 rpm for 6 hours and 14 minutes.

Forensic Science 12

Introduction

In almost all crimes of violence, whether murder, rape, or assault, there is usually some type of biological material left at the scene or on the victim. This material may be hair, saliva, sperm, skin, or blood. All of these contain DNA–DNA that may belong to the victim or to the perpetrator. The application of PCR to these samples has revolutionized law enforcement's capabilities in providing evidence to the courtroom where a suspect's guilt or innocence is argued.*

In forensics, PCR is used for the amplification of **polymorphic** sites, those regions on DNA that are variable among people. Some polymorphisms result from specific point mutations in the DNA, others from the addition or deletion of repeat units. A number of polymorphic sites have been identified and exploited for human identification. There are also a number of different techniques by which PCR products can be assayed and polymorphisms identified.

If the suspect's DNA profile and that of the DNA recovered from the crime scene do not match, then it can be said that the suspect is excluded as a contributor of the sample. If there is a match, then the suspect is included as a possible contributor of DNA evidence, and it then becomes important to determine the significance of the match. A lawyer involved in the case will want to know the probability that someone other than the suspect left the DNA evidence at the scene. The question is one of population statistics, and that is the primary focus of this chapter.

Alleles and Genotypes

An **allele** is one of two or more alternate forms of a genetic site (or **locus**). Some sites may have many different alleles. This is particularly true for sites carrying tandem repeats, where the DNA segment may be repeated a number of times, each different repeat number representing a different allele designation.

*The PCR process is covered by patents owned by Roche Molecular Systems, Inc. and F. Hoffmann-La Roche Ltd.

A **genotype** is an organism's particular combination of alleles. For any one polymorphic site, an individual will carry two possible alleles – one obtained from the mother, the other from the father. For an entire population, however, any possible combination of alleles is possible for the individuals within that population. For a locus having three possible alleles, *A*, *B*, and *C*, there are six possible genotypes. An individual may have any of the following allele combinations:

$$AA$$
$$AB$$
$$AC$$
$$BB$$
$$BC$$
$$CC$$

The number of possible genotypes can be calculated in two ways. It can be calculated by adding the number of possible alleles with each positive integer below that number. For example, for a locus having three possible alleles, the number of possible genotypes is

$$3 + 2 + 1 = 6$$

Likewise, for a five-allele system, there are 15 possible genotypes:

$$5 + 4 + 3 + 2 + 1 = 15$$

The number of genotypes can also be calculated using the equation

$$\text{number of genotypes} = \frac{n(n+1)}{2}$$

where n is equal to the number of possible alleles.

The first method is easier to remember but becomes cumbersome for systems having a large number of alleles.

Problem 12.1 A VNTR (variable number of tandem repeats) locus has 14 different alleles. How many genotypes are possible in a population for this VNTR?

Solution 12.1 Using the preceding equation, we have

$$\text{number of genotypes} = \frac{14(14+1)}{2} = \frac{(14)(15)}{2} = \frac{210}{2} = 105$$

Therefore, for a locus having 14 different alleles, 105 different genotypes are possible.

Calculating Genotype Frequencies

A **genotype frequency** is the proportion of the total number of people represented by a single genotype. For example, if the genotype *AA* (for a locus having three different alleles) is found to be present in six people out of 200 sampled, the genotype frequency is 6/200 = 0.03. The sum of all genotype frequencies for a single locus should be equal to 1.0. If it is not, the frequency calculations should be checked again.

Problem 12.2 A locus has three possible alleles, A, B, and C. The genotypes of 200 people chosen at random are determined for the different allele combinations, with the following results:

Genotype	Number of Individuals
AA	6
AB	34
AC	46
BB	12
BC	60
CC	42
Total	200

What are the genotype frequencies?

Solution 12.2 Genotype frequencies are obtained by dividing the number of individuals having each genotype by the total number of people sampled. Genotype frequencies for each genotype are given in the right column of the following table.

Genotype	Number of Individuals	Ratio of Individuals to Total	Genotype Frequency
AA	6	6/200 =	0.03
AB	34	34/200 =	0.17
AC	46	46/200 =	0.23
BB	12	12/200 =	0.06
BC	60	60/200 =	0.30
CC	42	42/200 =	0.21
Total	**200**		**1.00**

Note that the genotype frequencies add up to 1.00.

Calculating Allele Frequencies

An **allele frequency** is the proportion of the total number of alleles in a population represented by a particular allele. For any polymorphism other than those present on the X or Y chromosome, each individual carries two alleles per locus, one inherited from the mother, the other from the father. Within a population, therefore, there are twice as many total alleles as there are people. When calculating the number of a particular allele, homozygous individuals each contribute two of that allele to the total number of that particular allele. Heterozygous individuals each contribute one of a particular allele to the total number of that allele. For example, if there are six individuals with the *AA* genotype, they contribute 12 *A* alleles. Thirty-four *AB* heterozygous individuals contribute a total of 34 *A* alleles and 34 *B* alleles to the total. The frequency of the *A* allele in a three-allele system is calculated as

$$A \text{ allele frequency} = \frac{\left[2(\# \text{ of } AA\text{'s}) + (\# \text{ of } AB\text{'s}) + (\# \text{ of } AC\text{'s})\right]}{2n}$$

where n is the number of people in the survey. The frequencies for the other alleles are calculated in a similar manner. The sum of all allele frequencies for a specific locus should equal 1.0.

Problem 12.3 What are the frequencies for the *A*, *B*, and *C* alleles described in Problem 12.2?

Solution 12.3 We will assign a frequency of p to the A allele, q to the B allele, and r to the C allele. We have the following genotype data.

Genotype	Number of Individuals
AA	6
AB	34
AC	46
BB	12
BC	60
CC	42
Total	200

The allele frequencies are then calculated as follows:

$$pA = \frac{[2(6)+34+46]}{2(200)} = \frac{92}{400} = 0.230$$

$$qB = \frac{[2(12)+34+60]}{2(200)} = \frac{118}{400} = 0.295$$

$$rC = \frac{[2(42)+46+60]}{2(200)} = \frac{190}{400} = 0.475$$

Therefore, the allele frequency of A is 0.230, the allele frequency of B is 0.295, and the allele frequency of C is 0.475. Note that the allele frequencies add up to 1.000 (0.230 + 0.295 + 0.475 = 1.000).

The Hardy–Weinberg Equation and Calculating Expected Genotype Frequencies

Gregor Mendel demonstrated by crossing pea plants with different characteristics that gametes combine randomly. He used a **Punnett square** to predict the outcome of any genetic cross. For example, if he crossed two plants both heterozygous for height, where "*T*" represents a dominant tall phenotype and "*t*" represents the recessive short phenotype, the Punnett square would have the following appearance:

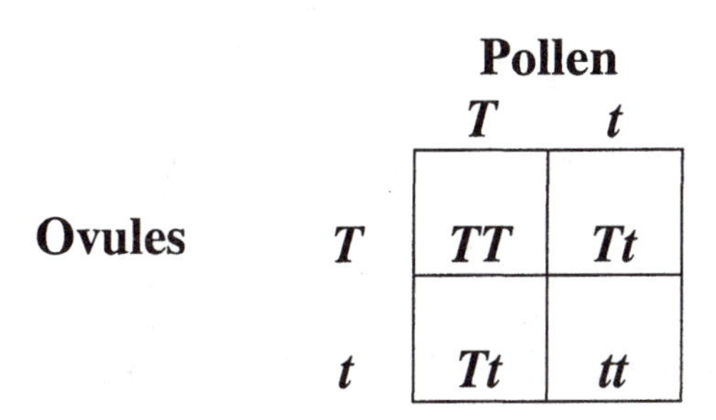

		Pollen	
		T	*t*
Ovules	*T*	*TT*	*Tt*
	t	*Tt*	*tt*

From the Punnett square, Mendel predicted that the offspring of the cross would have a phenotypic ratio of tall to short plants of 3:1.

G.H. Hardy, a British mathematician, and W. Weinberg, a German physician, realized that they could apply a similar approach to predicting the outcome of random mating, not just for an individual cross but for crosses occurring within an entire population. After all, random combination of gametes, as studied by Mendel for individual crosses, is quite similar in concept to random mating of genotypes. In an individual cross, it is a matter of chance which sperm will combine with which egg. In an infinitely large, randomly mating population, it is a matter of chance which genotypes will combine. Determining the distribution of genotype frequencies in a randomly mating population can also be accomplished using a Punnett square. However, rather than a single cross between two parents, Hardy and Weinberg examined crosses between all mothers and all fathers in a population.

For a locus having two alleles, *A* and *B*, a Punnett square can be constructed such that an allele frequency of p is assigned to the *A* allele and an allele frequency of q is assigned to the *B* allele. The allele frequencies are multiplied as shown in the following Punnett square:

		Fathers	
		pA	***qB***
Mothers	***pA***	***pApA***	***pAqB***
	qB	***pAqB***	***qBqB***

If *AA* represents the homozygous condition, then its frequency in the population is p^2 ($p \times p = p^2$). The frequency of the *BB* homozygous individuals is q^2. Since heterozygous individuals can arise in two ways, by receiving alternate alleles from either parent, the frequency of the AB genotype in the population is $2pq$ since $(p \times q) + (p \times q) = 2pq$.

The sum of genotype frequencies is given by the expression

$$p^2 + 2pq + q^2 = 1.0$$

This relationship is often referred to as the **Hardy–Weinberg equation**. It is used to determine expected genotype frequencies from allele frequencies.

Punnett squares can be used to examine any number of alleles. For example, a system having three alleles A, B, and C with allele frequencies p, q, and r, respectively, will have the following Punnett square:

		Fathers	
	pA	***qB***	***rC***
Mothers ***pA***	***pApA***	***pAqB***	***pArC***
qB	***pAqB***	***qBqB***	***qBrC***
rC	***pArC***	***qBrC***	***rCrC***

The sum of genotype frequencies for this three-allele system is

$$p^2 + q^2 + r^2 + 2pq + 2pr + 2qr = 1.0$$

This demonstrates that no matter how many alleles are being examined, the frequency of the homozygous condition is always the square of the allele frequency (p^2), and the frequency of the heterozygous condition is always two times the allele frequency of one allele times the frequency of the other allele ($2pq$).

The Hardy–Weinberg equation was derived for a strict set of conditions. It assumes that the population is in **equilibrium** – that it experiences no net change in allele frequencies over time. To reach equilibrium, the following conditions must be met:

1. Random mating
2. An infinitely large population
3. No mutation
4. No migration into or out of the population
5. No selection for genotypes (all genotypes are equally viable and equally fertile)

In reality, of course, there are no populations that meet these requirements. Nevertheless, most reasonably large populations approximate these conditions to an extent that the Hardy–Weinberg equation can be applied to estimate genotype frequencies.

Problem 12.4 What are the expected genotype frequencies for the alleles described in Problem 12.3? (Assume the population is in Hardy–Weinberg equilibrium.)

Solution 12.4 From Problem 12.3, we have the following allele frequencies:

$$pA = 0.230$$
$$qB = 0.295$$
$$rC = 0.475$$

The Hardy–Weinberg equation gives the following values:

Genotype	**Multiplication**	**Mathematics**	**Expected Genotype Frequency**
AA	(freq. of *A*)^2	$(0.230)^2 =$	0.053
AB	2(freq. of *A*)(freq. of *B*)	2(0.230)(0.295) =	0.136
AC	2(freq. of *A*)(freq. of *C*)	2(0.230)(0.475) =	0.218
BB	(freq. of *B*)^2	$(0.295)^2 =$	0.087
BC	2(freq. of *B*)(freq. of *C*)	2(0.295)(0.475) =	0.280
CC	(freq. of *C*)^2	$(0.475)^2 =$	0.226
Total			1.000

Note that the total of all genotype frequencies is equal to 1.000, as would be expected if the mathematics were performed correctly.

ıe Chi-Square Test: Comparing Observed to Expected Values

In Problem 12.4, genotype frequencies were calculated from allele frequencies using the Hardy–Weinberg equation. These genotype frequencies represent what should be expected in the sampled population if that population is in Hardy–Weinberg equilibrium. In Problem 12.2, the observed genotype frequencies found within that population were calculated. It can now be asked: Do the observed genotype frequencies differ significantly from the expected genotype frequencies? This question can be addressed using a statistical test called chi-square.

Chi-square is used to test whether or not some observed distributional outcome fits an expected pattern. Since it is unlikely that the observed genotype frequencies will be exactly as predicted by the Hardy–Weinberg equation, it is important to look at the nature of the differences between the observed and expected values and to make a judgment as to the "goodness of fit" between them. In the chi-square test, the expected value is subtracted from the observed value in each category, and this value is then squared. Each squared value is then weighted by dividing it by the expected value for that category. The sum of these squared and weighted values, called chi-square (denoted as χ^2), is represented by the following equation:

$$\chi^2 = \sum \frac{(\text{observed} - \text{expected})^2}{\text{expected}}$$

In the chi-square test, two hypotheses are tested. The **null hypothesis** (H_0) states that there is no difference between the two observed and expected values; they are statistically the same and any difference that may be detected is due to chance. The alternative hypothesis (H_a) states that the two sets of data, the observed and expected values, are different; the difference is statistically significant and must be due to some reason other than chance.

In populations, a normal distribution of values around a consensus frequency should be represented by a bell-shaped curve. The degree of distribution represented by a curve is associated with a value termed the **degrees of freedom** (*df*). The smaller the *df* value (the smaller the sample size), the larger is the dispersion in the distribution, and a bell-shaped curve will be more difficult to distinguish. The larger the *df* value (the larger the sample size), the closer the distribution will come to a normal, bell-shaped curve. For most problems in forensics, the degrees of freedom is equal to one less than the number of categories in the distribution. For such problems, the number of categories is equal to the number of different alleles (or the number of different genotypes, depending upon which is the object of the test). Statistically significant differences may be observed in chi-square if the sample size is small (less than 100 individuals sampled from a population) or if there are drastic deviations from Hardy–Weinberg equilibrium.

A chi-square test is performed at a certain level of significance, usually 5% ($\alpha = 0.05$; $p = 0.95$). At a 5% significance level, we are saying that there is less than a 5% chance that the null hypothesis will be rejected even though it is true. Conversely, there is a 95% chance that the null hypothesis is correct, that there is no difference between the observed and expected values. A chi-square distribution table (see Table 12.1) will provide the expected chi-square value for a given probability and *df* value. For example, suppose a chi-square value of 2.3 is obtained for some data set having 8 degrees of freedom and we wish to test at a 5% significance level. From the chi-square distribution table, we find a value of 15.51 for a probability value of 0.95 and 8 degrees of freedom. Therefore, there is a probability of 0.95 (95%) that the chi-square will be less than 15.51. Since the chi-square value of 2.3 is less than 15.51, the null hypothesis is not rejected; the data does not provide sufficient evidence to conclude that the observed data differs from the expected.

The chi-square test can be used in forensics to compare observed genotype frequencies with those expected from Hardy–Weinberg analysis and to compare local allele frequencies (observed) with those maintained in a larger or national database (expected values).

df	*p* = 0.75	*p* = 0.90	*p* = 0.95	*p* = 0.975	*p* = 0.99
1	1.32	2.71	3.84	5.02	6.64
2	2.77	4.60	5.99	7.37	9.21
3	4.10	6.24	7.80	9.33	11.31
4	5.38	7.77	9.48	11.14	13.27
5	6.62	9.23	11.07	12.83	15.08
6	7.84	10.64	12.59	14.44	16.81
7	9.04	12.02	14.07	16.01	18.48
8	10.22	13.36	15.51	17.54	20.09
9	11.39	14.68	16.92	19.02	21.67
10	12.5	15.9	18.3	20.5	23.2
11	13.7	17.3	19.7	21.9	24.7
12	14.8	18.6	21.0	23.3	26.2
13	16.0	19.8	22.4	24.7	27.7
14	17.1	21.1	23.7	26.1	29.1
15	18.2	22.3	25.0	27.5	30.6
16	19.4	23.5	26.3	28.8	32.0
17	20.5	24.8	27.6	30.2	33.4
18	21.6	26.0	28.9	31.5	34.8
19	22.7	27.2	30.1	32.9	36.2
20	23.8	28.4	31.4	34.2	37.6
21	24.9	29.6	32.7	35.5	38.9
22	26.0	30.8	33.9	36.8	40.3
23	27.1	32.0	35.2	38.1	41.6
24	28.2	33.2	36.4	39.4	43.0
25	29.3	34.4	37.7	40.7	44.3
30	34.8	40.3	43.8	47.0	50.9
40	45.6	51.8	55.7	59.3	63.7
50	56.3	63.2	67.5	71.4	76.2
60	67.0	74.4	79.1	83.3	88.4
70	77.6	85.5	90.5	95.0	100.4
80	88.1	96.6	101.9	106.6	112.3
90	98.7	107.6	113.2	118.1	124.1
100	109.1	118.5	124.3	129.6	135.8

Table 12.1 The chi-square distribution table. At 8 degrees of freedom, for example, there is a probability of 0.95 (95%) that the chi-square value will be less than 15.51. If the chi-square value you obtain is less than 15.51, then there is insufficient evidence to conclude that the observed values differ from the expected values. Below 15.51, any differences between the observed and expected values are merely due to chance.

Problem 12.5 In Problems 12.2 and 12.4, observed and expected genotype frequencies were calculated for a fictitious population. These are as follows:

Genotype	Observed Frequency	Expected Frequency
AA	0.03	0.053
AB	0.17	0.136
AC	0.23	0.218
BB	0.06	0.087
BC	0.30	0.280
CC	0.21	0.226

Using the chi-square test at a 5% level of significance (95% probability), determine if the observed genotype frequencies are significantly different from the expected genotype frequencies.

Solution 12.5 A table is prepared so that the steps involved in obtaining the chi-square value can be followed. So that we are not dealing with small decimal numbers, the frequencies given in the table will each be multiplied by 100 to give a percent value.

Genotype	Observed Frequency (O)	Expected Frequency (E)	Difference $(O-E)$	Square of Difference $(O-E)^2$	$(O-E)^2/E$
AA	3	5.3	−2.3	5.3	1.00
AB	17	13.6	3.4	11.6	0.85
AC	23	21.8	1.2	1.4	0.07
BB	6	8.7	−2.7	7.3	0.84
BC	30	28	2.0	4	0.14
CC	21	22.6	−1.6	2.6	0.11
Total					χ^2 = **3.01**

Therefore, a chi-square value of 3.01 is obtained. The degrees of freedom is equal to the number of categories (6, because there are six genotypes) minus 1:

$$df = 6 - 1 = 5$$

At 5 degrees of freedom and 0.95 probability, a chi-square value of 11.07 is obtained from Table 12.1. Therefore, there is a probability of 95% that χ^2 will be less than 11.07 if the null hypothesis is true, if there is not a statistically significant difference between the observed and expected genotype frequencies. Since our derived chi-square value of 3.01 is less than 11.07, we do not reject the null hypothesis; there is insufficient evidence to conclude that the observed genotype frequencies differ from the expected genotype frequencies. This suggests that our population may be in Hardy–Weinberg equilibrium.

Sample Variance

Sample variance (s^2) is a measure of the degree to which the numbers in a list are spread out. If the numbers in a list are all close to the expected values, the variance will be small. If they are far away, the variance will be large. Sample variance is given by the equation

$$s^2 = \frac{\sum (O-E)^2}{n-1}$$

where n is the number of categories.

A sample variance calculation is at its most powerful when it is used to compare two different sets of data, say genotype frequency distributions in New York City as compared with those in Santa Fe, New Mexico. A population that is not in Hardy–Weinberg equilibrium is more likely to have a larger variance in genotype frequencies than one that is in Hardy–Weinberg equilibrium. Populations far from Hardy–Weinberg equilibrium may have a sample variance value in the hundreds.

Problem 12.6 What is the sample variance for the data in Problem 12.5?

Solution 12.6 The sum the $(O-E)^2$ column in Problem 12.5 is

$$5.3 + 11.6 + 1.4 + 7.3 + 4 + 2.6 = 32.2$$

The value for n in this problem is 6 since there are six possible genotypes. Placing these values into the equation for sample variance gives the following result:

$$s^2 = \frac{32.2}{6-1} = \frac{32.2}{5} = 6.44$$

This value is relatively small, indicating that the observed genotype frequencies do not vary much from the expected values.

Sample Standard Deviation

The **sample standard deviation** (s) is the square root of the sample variance and is also a measure of the spread from the expected values. In its simplest terms, it can be thought of as the average distance of the observed data from the expected values. It is given by the formula

$$s = \sqrt{\frac{\sum (O - E)^2}{n-1}}$$

If there is little variation in the observed data from the expected values, the standard deviation will be small. If there is a large amount of variation in the data, then, on average, the data values will be far from the mean and the standard deviation will be large.

Problem 12.7 What is the standard deviation for the data set in Problem 12.5?

Solution 12.7 In Problem 12.6, it was determined that the variance is 6.44. The standard deviation is the square root of this number:

$$s = \sqrt{6.44} = 2.54$$

Therefore, the standard deviation is 2.54. The average observed genotype frequency is within 1 standard deviation (2.54 percentage points) from each expected genotype frequency. The AB and BB genotype frequencies are greater than 1 standard deviation from the expected frequencies (AB is 3.4 percentage points away from the expected value. BB is 2.7 percentage points away from the expected value; see Solution 12.5).

P_i: The Power of Inclusion

The **power of inclusion** (***P_i***) gives a probability that if two individuals are chosen at random from a population, they will have the same genotype. The smaller the P_i value, the more powerful is the genotyping method for differentiating individuals. The power of inclusion is equivalent to the sum of the squares of the expected genotype frequencies. For example, a P_i value of 0.06 means that 6% ($0.06 \times 100 = 6\%$) of the time, if two individuals are chosen at random from the population, they will have the same genotype.

Problem 12.8 What is the power of inclusion for the genotypes in Problem 12.4?

Solution 12.8 The expected genotype frequencies for the fictitious population are as follows:

Genotype	Expected Frequency
AA	0.053
AB	0.136
AC	0.218
BB	0.087
BC	0.280
CC	0.226

P_i is the sum of squares of the expected genotype frequencies. These are calculated as follows:

$$P_i = (\text{exp. freq. of } AA)^2 + (\text{exp. freq. of } AB)^2 + (\text{exp. freq. of } AC)^2 + (\text{exp. freq. of } BB)^2 + (\text{exp. freq. of } BC)^2 + (\text{exp. freq. of } CC)^2$$

Placing the appropriate values into this equation gives the following result:

$$P_i = (0.053)^2 + (0.136)^2 + (0.218)^2 + (0.087)^2 + (0.280)^2 + (0.226)^2$$

$$= 0.003 + 0.018 + 0.048 + 0.008 + 0.078 + 0.051$$

$$= 0.206$$

Therefore, 20.6% of the time, if two people are chosen at random from the population, they will have the same genotype.

P_d: The Power of Discrimination

The **power of discrimination** (**P_d**) is equivalent to 1 minus the P_i:

$$P_d = 1 - P_i$$

P_d gives a measure of how likely it is that two individuals will have different genotypes. For example, if a DNA typing system has a P_d of 0.9997, it means that 99.97% of the time, if two individuals are chosen at random from a population, they will have different genotypes that can be discriminated by the test.

A **combined P_d** is equivalent to 1 minus the product of P_i of all markers. It is used when multiple loci are being used in the typing assay. The P_i value for a single genetic marker is equivalent to 1 minus the P_d value for that marker. The more loci used in a DNA typing test, the higher will be the combined P_d.

Problem 12.9 What is the power of discrimination for the genotypes in Problem 12.8?

Solution 12.9 In Problem 12.8, a P_i value of 0.206 was calculated. P_d is 1 minus P_i. This yields

$$P_d = 1 - 0.206 = 0.794$$

Therefore, 79.4% of the time, if two individuals are chosen at random from a population, they will have different genotypes at that locus.

DNA Typing and a Weighted Average

A genotype frequency is derived by using the Hardy–Weinberg equation and observed allele frequencies. Many crime laboratories obtain allele frequency values from national databases. These databases may have allele frequencies for the major population groups (Caucasian, African American, and Hispanic).

A crime occurring within a city is usually perpetrated by one of its inhabitants. For this reason, most major metropolitan areas will develop an allele frequency database from a random sample of its own population. Even without this data, however, a city can still derive a fairly good estimate of frequencies for the genotypes within its boundaries. This can be done using census data. For example, the proportion of Caucasians within the city's census area can be multiplied by a national database genotype frequency to give a weighted average for that population group. The weighted averages for each population group are then added to give an overall estimate of the frequency of a particular genotype within the census area.

Problem 12.10 A blood spot is typed for a single STR (short tandem repeat) locus. It is found that the DNA contains five repeats and six repeats (designated 5, 6, read "five comma six") for this particular STR. The FBI database shows that this heterozygous genotype has a frequency of 0.003 in Caucasians, 0.002 in African Americans, and 0.014 in Hispanic populations. The city in which this evidence was collected and typed has a population profile of 60% Caucasians, 30% African Americans, and 10% Hispanic. What is the overall weighted average frequency for this genotype?

Solution 12.10 Each genotype frequency will be multiplied by the fraction of the population made up by each ethnic group. These values are then added together to give the weighted average. The calculation is as follows:

$$\begin{aligned}\text{weighted average} &= (0.60 \times 0.003) + (0.30 \times 0.002) + (0.10 \times 0.014) \\ &= 0.0018 + 0.0006 + 0.0014 = 0.0038\end{aligned}$$

Therefore, the weighted average genotype frequency for a 5, 6 STR type for the city's population is 0.0038. This can be converted to a "1 in" number by dividing 1 by 0.0038:

$$\frac{1}{0.0038} = 263$$

Therefore, 1 in approximately 263 randomly chosen individuals in the city would be expected to have the 5, 6 STR genotype.

The Multiplication Rule

The more loci used for DNA typing, the better the typing is able to distinguish between individuals. Most DNAs are typed for multiple markers, and then the genotype frequencies are multiplied together to give the frequency of the entire profile. This is called the **multiplication rule**. For its use to be valid, the loci must be in linkage equilibrium. That is, the markers are on different chromosomes or far enough apart on the same chromosome that they segregate independently; they are not linked.

Problem 12.11 A DNA is typed for five different, unlinked loci. The following genotype frequencies for each locus are determined:

Locus	Genotype Frequency
1	0.021
2	0.003
3	0.014
4	0.001
5	0.052

What is the genotype frequency for the entire profile?

Solution 12.11 To determine the genotype frequency for the complete profile, the genotype frequencies are multiplied together:

$$\text{Overall frequency} = 0.021 \times 0.003 \times 0.014 \times 0.001 \times 0.052 = 4.6 \times 10^{-11}$$

Therefore, the genotype frequency for the entire profile is 4.6×10^{-11}. This can be converted to a "1 in" number by dividing 1 by 4.6×10^{-11}:

$$\frac{1}{4.6\times10^{-11}} = 2.2\times10^{10}$$

Therefore, from a randomly sampled population, 1 in 2.2×10^{10} individuals might be expected to have this particular genotype combination.

Index

have seen, the underlying sentiment was nothing exceptional in Roman society. Female modesty had always been appreciated. The passages do not reflect new attitudes towards women, but they show that Constantine's chancellery displayed an active interest in inherited moral values, perhaps more so than governments before and after him. It is a pity that we do not know who drafted his laws.

Although women were not forbidden to plead their case in person, they were hardly an everyday sight in Roman courts. A respectable matron was expected to acquire a male representative who could appear before the court either on her behalf or together with her. On the other hand, women's personal participation was clearly not exceptional either.[39] From the late third century to the sixth the presence of fourteen female litigants is attested in papyri. Within the same period ten women are represented by males.[40]

Little certain can be said about later developments in the west. It is unlikely that outright prohibitions ever surfaced in imperial law. A few Merovingian *formulae* indicate that women could still be present during litigation. Formal rules hardly played a noteworthy role in early medieval jurisdiction. Those who nursed a grievance against someone appealed to bishops or the king himself. Females were lucky to have a man to speak on their behalf but that was no prerequisite.[41]

It is difficult to form a comprehensive picture of women's public appearance in the classical world. The Romans could occasionally claim that keeping women out of the public eye was something typical of Greeks. There were indeed regions in the eastern Mediterranean where women went veiled outside their home, but this was by no means a general rule, and even some Roman males would not have thought it was such a bad idea. They always believed that women had behaved themselves better in the good old days, whereas contemporary women were

[39] See e.g. Val. Max. 8. 3. 1–2; Ulp. D. 26. 10. 1. 7; CI 2. 12. 2 (161), 2. 12. 4 (207), 2. 9. 3 (294); FV 326 (294?); CT 8. 15. 1 (Constantine); Ambr. *in psalm. 118 serm.* 16. 7. 2; Gardner (1986) 262–3; Marshall (1989), esp. 50; Krause (1994*b*) 244–51.

[40] Beaucamp (1992*a*) 21–8, cf. also 275. On the early Roman period and on P.Mich. viii. 507 (a private letter, mistakenly claiming that women needed a representative), see Anagnostou-Cañas (1984) 358–9.

[41] *F.Andec.* 12, 16; *F.Sal.Bign.* 7; cf. also LRib 84; Greg. Tur. *franc.* 9. 33; *Conc. Matisc.* (585) 12, CCL 148A. 244.

far too independent. The familiar terms *pudicitia* and *verecundia* appear frequently in Latin literature, though describing less actual feminine behaviour than a lost ideal. In fact, it is not at all clear that women in the Greek east generally lived a more restricted life than their sisters in the west. On the contrary, the social 'emancipation' of women may have started much earlier in the prosperous hellenistic urban culture.[42]

Be this as it may, women's public behaviour was not first and foremost a legal issue. Only two facts might be stressed here. First, we have the hundreds of imperial rescripts which Diocletian addressed to women: these women mostly lived in the eastern half of the empire. True, petitions did not necessarily presuppose a personal encounter with the emperor or active litigation in court, nor do we know how many petitions the western emperors received from women. But it is remarkable that almost one-third of all eastern rescripts in the 280s and 290s had female addressees. This bare fact testifies against any general, or at least successful, women-keep-silent-at-home policy.

Secondly, whatever the public role of women in the late Roman east, it certainly did not have any effect on Constantine's legislation. Those three laws which stressed feminine modesty were all written in the first decade of his reign. At that time he ruled only over the western half of the Roman empire, and his chancellery hardly knew anything about conditions in the east. We can suspect considerable local and social variations in women's public image, but the existing evidence does not speak for any large-scale geographic patterns.

Nor can we demonstrate changes over the course of time. Perhaps the most notable development was the emergence of the Christian ascetic ideal. It demanded a very strict standard of behaviour from monastic women: in extreme instances they were expected to veil their face in public (Hier. *ep.* 22. 27; 130. 18). On the other hand, we have to remember that tender virgins had

[42] See e.g. Nep. praef.; Philo Alex. *leg. spec.* 3. 169–77; Hor. *carm.* 3. 6. 24; Liv. 34. 1–8; Val. Max. 3. 8. 6, 6. 3. 10–12; Colum. 12. praef. 7–10; Sen. *benef.* 1. 9. 3; Tac. *ann.* 3. 33–4; Suet. *Aug.* 44; *Tib.* 50; Juv. 6. 286–300; Plut. *conjug. praec.* 30–2; Tert. *apol.* 6. 6, CCL 1. 97; cf. further L. Robert, *Hellenica*, 5 (Paris, 1948), 66–9; Thraede (1972) 197–204, 218; MacMullen (1980), for the veil esp. 208–9, 217–18; Van Bremen (1983); Treggiari (1991*a*) 211–15; Dixon (1992) 19–24; Gardner (1993) 101–7; Saller (1994) 5.

always been guarded closely in the Mediterranean world, without any religious motives (e.g. Philo Alex. *leg. spec.* 3. 169). As far as ordinary married women were concerned, it is unlikely that Christian teaching meant anything new for their lifestyles. They were advised to avoid publicity and to take care of domestic matters, such as the proverbial working of wool (*lanificium*). Nothing indicates that this was recognized as peculiarly Christian or that the question aroused more than casual interest among the bishops. Whether the ideal was realized better in late antiquity than in the principate remains in doubt. The letters of Jerome, for example, suggest that in late fourth-century Rome both younger and older women appeared freely in public without the slightest scruples.[43]

Here we may also recall the famous female philosopher Hypatia who had a prominent position and high connections in Alexandria. She gave public lectures on Platonism and mathematics. We can never know the actual reasons why she was lynched by a Christian mob in 415 but it was hardly because she was a woman and behaved improperly.[44]

Christian women themselves did not even stay within the walls of their home towns. Pilgrimages became so fashionable in the fourth and fifth centuries that it can be compared to modern tourism. Innumerable women from all over the empire travelled to Syria, Palestine, and Egypt to view the holy places and the arresting communities of monks. They may not actually have journeyed alone: aristocratic ladies were often followed by an entourage of eunuchs, slaves, and clerics. There was an enormous difference between a pious nun's humble company and the regal pomp of some less devout travellers. In any case, the mere existence of this phenomenon shows that women in late antiquity were not doomed to invisibility or idleness.[45]

[43] See for the ideal of quiet domestic life, e.g. Auson. *parent.* 2; 16. 4 (= 161, 175); Jul. *elog. Euseb.* 127–8; Greg. Naz. *carm.* 2. 2. 6, PG 37. 1542ff.; Symm. *ep.* 6. 67; Joh. Chrys. *propter forn.* 2; *qual. duc. ux.* 7, PG 51. 211, 236; *in psalm.* 48. 7, PG 55. 510. For women in public, Hier. *ep.* 77. 4, with other references above in Ch. 5, Sect. 1; Sidon. *ep.* 5. 17. 3; Greg. Tur. *franc.* 7. 8; Greg. M. *ep.* 4. 9. Cf. further Thraede (1972) 215–16, 239–52; MacMullen (1986*b*) 326–8; Brown (1988) 205–8; Treggiari (1991*a*) 414–27; Beaucamp (1992a) 289–91, 348–9.

[44] On Hypatia, see Rist (1965).

[45] For pilgrimage, see Hunt (1982) *passim*; e.g. Pall. *hist. laus.* 35, 46; Hier. *ep.* 54. 13, 108. 6ff.; Paul. Nol. *ep.* 29. 10, CSEL 29. 257; Geront. *vita Melaniae* 34–9, 50, 56, SC 90. 191ff.; and, of course, the *Itinerarium Egeriae*.

Professions, Honours, and Obligations

When classical scholars became interested in the female half of the ancient world they posed the simple question: what did the silent women of Greece and Rome do while their menfolk wrote epic poems or killed each other on the battlefield? The legal evidence provides at least one answer: they counterfeited money. So much appears from a law of Constantine which states that counterfeiters were punished 'according to their sex and social standing' (CT 9. 21. 1).

This is a mere curiosity, of course. Legal sources rarely mention women's real 'professional' activities. Much of the ancient production and service industries depended on the use of slaves. For example, the mass production of textiles in state-owned factories in late antiquity was largely based on unfree labour, both male and female.[46] There were occupations for freedwomen and poor freeborn women, too. It is well known that they practised many types of professions, from cloth-making and retail trade to midwifery and prostitution. Especially the production of clothing is frequently connected with females. However, working women are only passingly mentioned in juristic writings.[47] Their business was far too insignificant. If a customer did not pay for his cloak, it was not worth the trouble to ask for an imperial rescript. And quite predictably, the general edicts of the later empire show still less concern for lower-class professional activities. Most people in the Roman empire transferred only once or twice in their lifetime such amounts of property that a legal process was worth while: inheritance and dowry. It is no wonder that precisely these issues are very common in the rescripts addressed to women.[48]

The legislators were mainly interested in those people who had inherited enough wealth to live off it. Most of the private capital was invested in arable land and the slaves cultivating it. Whether rich senators or humble members of the curial class, the owners received their income from the selling of agricultural products. Women were naturally included in these sections of the society.

[46] CT 10. 20. 2–3; Demandt (1989) 341–2; cf. also other laws of CT 10. 20.

[47] D. 14. 3. 7–8, 15. 1. 27. pr, 32. 65. 3, 34. 2. 32. 4, 37. 15. 11, 50. 13. 1. 2 + 14; PS 2. 8. 1. For the range of female professions, see e.g. Treggiari (1979); Gardner (1986) 237–53; Herlihy (1990) 1–44; Krause (1994*b*) 130–60; and for slaves, Treggiari (1976).

[48] Cf. Huchthausen (1974) 214–26; Sternberg (1985) 514–16, 520–5.